PIVOTAL STRATEGIES

Pivotal Strategies

Claiming Writing Studies as Discipline

EDITED BY
Lynn C. Lewis

UTAH STATE UNIVERSITY PRESS
Logan

© 2024 by University Press of Colorado

Published by Utah State University Press
An imprint of University Press of Colorado
1580 North Logan Street, Suite 660
PMB 39883
Denver, Colorado 80203-1942

All rights reserved

 The University Press of Colorado is a proud member of Association of University Presses.

The University Press of Colorado is a cooperative publishing enterprise supported, in part, by Adams State University, Colorado State University, Fort Lewis College, Metropolitan State University of Denver, University of Alaska Fairbanks, University of Colorado, University of Denver, University of Northern Colorado, University of Wyoming, Utah State University, and Western Colorado University.

ISBN: 978-1-64642-631-7 (hardcover)
ISBN: 978-1-64642-632-4 (paperback)
ISBN: 978-1-64642-633-1 (ebook)
https://doi.org/10.7330/9781646426331

Library of Congress Cataloging-in-Publication Data

Names: Lewis, Lynn (Lynn C.), editor.
Title: Pivotal strategies : claiming writing studies as discipline / edited by Lynn C. Lewis.
Description: Logan : Utah State University Press, [2024] | Includes bibliographical references and index.
Identifiers: LCCN 2024002452 (print) | LCCN 2024002453 (ebook) | ISBN 9781646426317 (hardcover) | ISBN 9781646426324 (paperback) | ISBN 9781646426331 (ebook)
Subjects: LCSH: English language—Rhetoric—Study and teaching (Higher) | English language—Composition and exercises—Study and teaching (Higher) | Report writing—Study and teaching (Higher) | Academic writing—Study and teaching (Higher) | Creative writing (Higher education)
Classification: LCC PE1404 .P55 2024 (print) | LCC PE1404 (ebook) | DDC 808/.0420711—dc23/eng/20240223
LC record available at https://lccn.loc.gov/2024002452
LC ebook record available at https://lccn.loc.gov/2024002453

Cover photo: © Shutterstock/Sven Hansche

For Peggy Anne Williams
Brave, Brilliant, Beautiful, Beloved.
1930–2019

Contents

Acknowledgments ix

Introduction
    Lynn C. Lewis 3

Section I: Kairos and Opportunity

    Interlude One: Kairos and Opportunity 21

1. Political, Personal, and Pedagogical Imperatives: Tactical Disciplinarity among Early Members of Writing Studies
Lauren Marshall Bowen and Laurie A. Pinkert 23

2. Through the Eyepiece (and Body) of a Long Past
Suellynn Duffey 44

3. Strategizing Disciplinarity, Disciplinary Strategies
Tara Wood 61

4. Embracing Failure: A Newly Independent Department's Attempts at Writing Its Own Script
Ron Brooks, Caroline Dadas, Laura Field, and Jessica Restaino 76

Section II: Negotiations and Resilience

    Interlude Two: Negotiations and Resilience 103

5. Claiming and Being Claimed by Writing Studies: Negotiating Identities for Creative Writers Teaching Composition
Alison Ersheid, Lisa Konigsberg, Maureen McVeigh, Nancy Pearson, and Seth Kahn 105

6. From Pell Grants to Tenure Track: Precarity and
   Privilege as a Disciplinary Pathway
   Cynthia Johnson   131

7. Finding Resilience in Writing Studies at the United States–Mexico Border
   Karen R. Tellez-Trujillo   148

Section III: Allegiance and Identification

Interlude Three: Allegiance and Identification   165

8. Being the Only One: The Embodiment and Labor of Tokenism
   Khadeidra Billingsley   167

9. Literacy and Disciplinarity: Vignettes of Struggle and Identification
   Raymond D. Rosas   182

10. Cognitive Dissonance
    Alison Wells Zepeda   195

11. Writing into Inclusion from the Margins
    Antonio Byrd   207

Section IV: Conclusion

Embodying Stories: Writing Studies and Its Potential Paths
Christina V. Cedillo   229

Index   239
About the Authors   255

Acknowledgments

Building a collection of essays from call for papers all the way to publication demands commitment, time, patience, and irrepressible belief in the importance of the stories that are being told. I acknowledge first the contributors to this book. The chapters in this book emerged from contributors facing down personal and professional challenges, tenure clocks ticking, and the COVID-19 miasma that took over all of our lives in one form or another. These contributors offer vulnerability in their stories while also insisting on their membership, fraught though it may be, in the discipline of writing studies.

I thank Dr. Christina V. Cedillo in particular for their support throughout this project. As a founding editor of the *Journal of Multimodal Rhetorics*, they gave me sound advice on offering supportive critique to contributors and, at my invitation, wrote the concluding chapter to this book. The book was made better because of them.

Thanks also go to Dr. Josh Daniel, with whom I worked collaboratively in the early stages of this project.

Dr. An Cheng and Dr. Bill Decker insisted I apply for sabbatical so that I could work without interruption on my writing, and An's thorough editorial eye ensured that I filled out the Request for Sabbatical form correctly and got it in on time. (This was a more difficult task than it sounds.) An and Bill have been consistently supportive of my research in all the best ways.

My thanks to Oklahoma State University and the English Department in particular for giving me a scholarly home for the past fourteen years and for its continual support of faculty research efforts.

Rachael Levay, former editor in chief of Utah State University Press / University of Press Colorado, possesses depth of knowledge and commitment to rhetoric and writing studies. She has encouraged me from the very beginning and was a creative and patient problem-solver and listener. She is an "author's editor," par excellence. Grateful thanks go to two anonymous reviewers for their careful, smart, and thoughtful suggestions; they truly made the book stronger.

My thanks also go to the brilliant and hard-working graduate students in the Rhetoric and Writing Studies program at Oklahoma State University. Jeaneen Canfield, Christina Lane, and Sarah Lonelodge, with whom I share a writing group, earned their PhDs in the dark COVID-19 spring of 2021 and have been sources of inspiration as I write. Fertile and smart conversations in my History of Rhetoric graduate seminar included Bernadette Bray, Gloria Evans, Elizabeth Key, Kevin Kourakos, Will Smith, and Richard Sylvestre. Their insights fed my imagination as I drew together the parts of this manuscript.

Finally, but never last, to my family. Micah and Megan give me joy, laughter, and happiness every day, and I am always proud and thankful as I watch them follow their paths. To Mitchell, my dearest, thank you for the perfectly brewed home-roasted coffee every morning and the warm constancy of love.

PIVOTAL STRATEGIES

PIVOTAL STRATEGIES

Introduction

LYNN C. LEWIS

On New Year's Day in January of 2022, I posted a question to my Facebook account: "Informal poll for people in the field of Rhetoric and Writing Studies: what metaphors have you/did you used/use for grad students who change disciplines, i.e., from literature to RWS or Creative Writing to RWS, etc.?"

Eighty-one comments later, I saw a picture emerging: The language used to describe a move to rhetoric and writing studies or composition and rhetoric frequently bore negative valence. A number of posters referred to "coming to the Dark Side" or "defecting" from, presumably, the good guys. Others remembered terms like "retread," "stepchild," and "bastard stepchild," a suggestion that our field lacks authenticity and leeches from more genuine work. One thread highlighted how the field of composition and rhetoric or writing studies was more frequently aligned with terms like "selling out" and "capitulating"—the suggestion here being that although one was more likely to be hired into a tenure-track job and to be able to make a living, this meant one was no longer doing the real, hard, intellectual work.

Three posters, however, made claim to neutral terms and argued, either implicitly or explicitly, that they had never been shamed for their choice of discipline nor had they ever felt demeaned. Fair enough. Still, the immediate consensus was that this question reflected experiences encapsulated in

https://doi.org/10.7330/9781646426331.c000

language choices that floated in the air of classrooms, graduate student coffee rooms, and cheap bars. For a long time, for most of us, choosing to identify with this "bastard stepchild" of a discipline has meant thinking about how we *differed* as well as who we are. We differed from literature because we cared more about writing and pedagogy; we differed from creative writing because what we wrote did not easily fit the designated creative genres. Every time I taught the required graduate student introduction to composition theory and pedagogy, I fielded questions about how writing studies differed from other disciplines. It is not a bad question. But focusing on differences elides the more substantive and well-theorized question of how writing studies became and maintains itself as a discipline.

The field's quest for disciplinarity has drawn the attention of numerous scholars. For example, the 2010 special issue of *CCC*, "The Future of Rhetoric and Composition," includes an important essay by Louise Phelps and John M. Ackerman in which they argue for the continuing need for scholars in composition and rhetoric/writing studies to make "the case for disciplinarity." Focusing on the *Visibility Project*, an initiative meant to ensure the recognition of composition, rhetoric, and writing studies in official databases, Phelps and Ackerman note that "to be recognized as a discipline is a powerful measure of whether we have earned the respect of others" (2010, 181). That recognition, as my informal experiment above demonstrates, continues to be an issue. We still seek respect.

"We" in these sentences calls for unpacking, however. Rhetoric and writing studies has, like many academic disciplines, suffered from terministic screens infused with whiteness. In fact, as Keith Gilyard, Adam Banks, Jacqueline Jones Royster, Susan Kates, and Sharon Crowley among many others have argued, racialized minority voices and bodies have been erased and elided. Nevertheless, their work as scholars and teachers has been integral to the construction of writing studies as discipline. As early as 1999, Gilyard called for continued rigorous attention to archival research in order to redress the problem of elided voices. He underscores his argument with a rich list of Black scholars whose work has been unrecognized but foundational to the field. Drawing on Susan Kates's work, he describes the pedagogy and writings of Haillie Quinn Brown, for instance, who in the early twentieth century developed pedagogical strategies with the then-radical assumption of developmental writing. Gilyard maps the work of Carter Woodson, W. E. B. Du Bois, and Geneva Smitherman, among others, describing how each scholar-teacher influenced the new discipline of composition and rhetoric. As Gilyard puts it,

"We may not always have been in the house of mainstream composition studies, but we were always knocking on the door" (1999, 642–643).

Knocking on closed doors comes with costs. Claiming the discipline as one's own can be costly on individual, community, and programmatic levels. For example, in his description of the effects of racial politics, Vershawn Ashanti Young has detailed the "rhetorical performance" necessary in negotiating his identity as a Black, male professor ("Your Average Nigga"). Stories about claiming a place in writing studies orbit questions of personal identity as well as professional and disciplinary belonging.

Racial politics and the ethics of racial encounters are crucial touchstones in our field now and they have taken on an urgency long overdue in the humanities in general. While this book does not take up the racial politics of our field explicitly, the contributors' stories here cannot help but be inflected by them. While the field has certainly orbited many other issues, problems, and questions, it has never listened hard enough to those knocks on the door.

I do not claim exceptionalism for writing studies in its complicated becoming. Challenges to disciplinarity have emerged in the humanities as well as the social sciences. Gary Olson has described anthropology's struggles over its identity. In writing studies, struggles over whether we are a service discipline or an intellectual one, whether theory matters when our focus is or should be pedagogy, and similar concerns can be tracked (Olson 2002). As Olson explains, "Anxieties of status and self-definition presented a common topic for discussion and research in the field." Put reductively, the *what* of writing studies was an early obsession for scholars. Olson's edited collection *Rhetoric and Composition as Intellectual Work* not only pushes back at the notion that service is primary but also suggests an early strategy for claiming writing studies as discipline: "The truth of the matter is that rhetoric and composition already is an intellectual discipline," says Olson (2002, xii). While aside Olson's main purpose for the collection, a book arguing for the intellectual worth and weight of its scholars is making a claim to disciplinarity and delineating its borders.

Defining the *what* of writing studies is a move common to the discipline; similar books and journal articles abound. As Heilker and Vandenberg (2015) show in their *Keywords in Writing Studies* collection, how borders have been or should be drawn about the nature of writing studies remains of great interest. Heilker and Vandenberg document the wonderful variety of border tracing; from Wardle and Downs's call for a writing-about-writing curriculum to Charles Bazerman's exuberant argument about the broad interdisciplinary reach of the field, the lines demarking writing studies are drawn and redrawn across the map of the humanities. While the redrawing of lines is not unique to writing studies, the field's rigorous interrogation of its existence, methods, purpose, and exigencies is certainly unusual.

But what approaches have scholars constructed in order to claim membership in the field? This is the *how* that concerns this book. That is, *how* individuals, programs, and departments strategize their becoming, identifying, and claiming. What are their key—pivotal—strategies? Answers to this question will require tracing the individual exigencies, purposes, rhetorical contexts, and kairotic moments leading to the construction and development of those strategies. They will require interrogation of the exclusion of marginalized voices and conscientious listening. The answers will require stories, in other words.

This collection turns on the notion that we tell ourselves stories in order to claim our discipline. Whether personal, social, communal, or institutional, these stories help us think about what constitutes writing studies and what deliberate, kairotic moves we writing studies scholars make. Only in the last five years or so have undergraduate writing studies majors been possible at some (but not all) universities. That means graduate students may not have arrived with the desire to pursue scholarship in composition and rhetoric or professional writing or writing studies. That means there must have been a shift in their thinking, a change in the questions that intrigued them, the scholars they wanted to read, and the expertise they sought to gain. These students had to locate themselves as writing studies scholars through conscious, purposeful moves. These are stories worth hearing. Likewise, in the 1980s as composition and rhetoric or writing studies emerged as a field or discipline or subject (the terms, while not fungible, shift from year to year), scholars trained in literature or creative writing or other fields reinvented themselves as they invested in the chief concerns of writing studies. Those stories, also, are worth hearing.

Indeed, the process of becoming requires a genuine pivot in thinking not least because there have been so few undergraduate degrees in writing studies until now. While the last five years have seen moderate growth in undergraduate degrees in writing studies, many writing programs and/or English departments still do not offer a Bachelor of Arts degree in writing studies. New graduate students often know little about rhetoric and writing studies when they begin their degree programs and may then unexpectedly find themselves pulled into questions about writing and literacy that they genuinely want to explore. It is a common phenomenon and a traceable through line among the stories in this book.

Claiming the discipline is not only about individual stories. The birth of standalone writing studies departments, for example, is a notable new phenomenon. What stories are there? As a discipline, we can be and often are lumped together with other "Englishy" fields. What negotiations, emergences, and challenges do writing studies programs face when they are not standalone? My own department houses five different programs: Creative Writing, Literature, Rhetoric and Writing Studies, TESL/Linguistics, and Screen Studies. The department has experienced challenges and occasional tensions around its multidisciplinary construction. As faculty at a Research 1 university, each of us are highly sensitive to the importance and impact of the scholarship and creative work we accomplish. We are also highly sensitive to questions of language and belonging as any multidisciplinary department would be.

I am standing at the front of the faculty meeting with a four-page handout. The handout details the widespread use of the term "writing studies" and I am making the argument that our program should be granted permission to change our name from Rhetoric and Professional Writing to Rhetoric and Writing Studies. The director of our program has just offered our rationale for the change, and it is my job to provide evidence that the term is widely used and widely accepted by people in our field.

This is our program's second attempt to make the argument. The first, an abysmal failure, was met with strong opposition from one of our department's five programs and mild puzzlement from the other three. We were directed to attempt to "work it out" with the program whose faculty were objecting. This, too, was an abysmal failure. So, after a year's delay and multiple attempts at negotiation, we are proposing the change again.

Our junior faculty are wary of making the argument once more. They are rightfully weary of what has to come to feel like an overly vitriolic debate since some are so strongly opposed to our change, and junior faculty, not yet tenured, feel targeted. But we all believe the change is essential; our current program name fails to capture what we really teach, research, and write about and is, for us, simply outdated.

As I stand in front of the faculty, I feel some confidence. Evidence of the widespread use of the term "writing studies" is present at peer institutions, in job ads, in CFPs, conference calls, recent book titles, and book series, so I have plenty to say. But as I walk faculty through the evidence, I also find myself feeling as if I have been reduced to bargaining for my discipline's identity, insisting on its significance, and asking permission to use the descriptors it has birthed. My pivotal strategy in this story was to locate and demonstrate the prevalence of the term "writing studies." While not particularly innovative, the strategy worked in this case because my audience was unaware of the term's prevalence, and I was able to show its acceptance across a range of academic writing artifacts familiar to them: for this rhetorical context and kairotic moment, the strategy was pivotal.

My approach was dependent on the labors of past scholars who struggled to name a discipline notable for its messy insistence on blurring boundaries. As early as 1989, writing program administrators (WPAs) Armstrong and Fontaine described their experiences, noting that "we feel that the creative power of naming then is particularly important in an emerging discipline such as Composition" (1989, 9). They note the challenge of making an informed decision about naming their own positions as WPAs, their staff and graduate student assistants, and even the content of their pedagogy. They also see the emergence of the discipline as demanding not only inquiring into the *what* of the field but also presenting an opportunity to name. Challenges necessitating negotiations permeate their stories. This is familiar still.

The question of naming and defining will likely persist especially as the field continues to evolve. Claire Lauer's (2012) "What's in a Name: The Anatomy of Defining New/Multi/Modal/Media Texts" reminds writing scholars that "defining terms . . . helps us discover what we value and where we stand in relation to what has been said and done before." Given the ever-changing nature of the field—despite efforts to "name what we know" or delineate "threshold concepts"—Lauer's point helps explain the field's necessary focus on this question.

Yet, as I think about the approaches our program took to persuade other faculty that our name change was reasonable, more questions emerge. What are the essential approaches to claiming the discipline? How do such claims work at programmatic, collective, and individual levels? Arguing for a particular name or disciplinarity requires strategic and tactical approaches. Similarly, claiming space or identity as a writing studies scholar demands exigency, agentive movement, and commitment. Given Gilyard's reminder of Black scholars knocking at the door, how should writing studies understand—and amplify—the experiences of underrepresented groups? Writing studies lauds its openness to racialized minorities, working class, and queer teacher-scholars, but what does it really mean to say "Yes, I belong *here*"?

In order to claim the discipline, whether as an individual, a program, or a department, two critical components must be in place. First, the field must convince outsiders that it exists and that its existence matters. Second, the field must achieve some insider consensus about its characteristics, methods, and values. On the other hand, consensus must account for the ever-shifting nature of writing studies and recognize that it is always contingent and always in need of renewal. Writing studies scholars have worked and continue to work at persuading outsiders and achieving consensus, if only briefly, within.

Louise Phelps's work with the Visibility Project exemplifies an approach to persuading outsiders. For Phelps, seeing that rhetoric and composition did not exist in numerical taxonomies of disciplines created by the National Research Council (NRC) and other groups meant "we had discovered a hidden mechanism that was playing a major role in keeping the field invisible." As she details in an interview, "Basically there is information disseminated about fields through statistics and databases that count and describe things like programs, degrees, and faculty, using discipline-based codes. As a field we weren't represented in the codes, not just in the NRC taxonomy but in an array of others. They reinforce each other because they're so interconnected" (Rodrigue 2013). Phelps's work to ensure composition and rhetoric was included was strategic and important to the discipline claiming a space for itself. Phelps, working with John Ackerman and the Doctoral Consortium, developed strategic approaches to making composition and rhetoric visible that included demonstrating that the number of students graduating with PhDs was comparable to already recognized and sustainable disciplines as well as surveying doctoral programs and gathering vast quantities of data. As Phelps and Ackerman explain when detailing two critical cases where rhetoric and writing studies was elided in the codes, "Each of these cases is a classic instance of rhetorical work. Each involved a complex collaborative process of research, data gathering, invention, and communication that was fraught with obstacles and difficulties" (2010, 186).

This work rendered the discipline visible because investigating and learning the complexities of institutional coding and then arguing for inclusion effected change at multiple levels on a broad scale. Departments and programs seeking to stake a claim can call on the codes to instantiate their arguments. The focus on outsider views threads through the stories in this book.

As the chapters in this collection will demonstrate, the question of naming and defining has given rise to a number of pivotal strategies as well as tactical approaches to the problem of claiming the discipline. However, the exigencies leading to the stories in these chapters vary in ways that reflect our cultural moment. Two particular exigencies shine here: the first lies in the overwhelming necessity of committing to antiracist practices and listening to the voices of BIPOC people. The second is the unsustainable and permeating sense of precarity in writing studies as well as other disciplines in the humanities.

The field's current focus on antiracist practices is evident in the flurry of position statements now appearing on the Conference on College Composition and Communication (CCCC) website, especially since 2016. These include statements on ensuring equitable hiring practice and a diverse candidate pool as well as numerous useful and important statements on language, such as the "Statement on Language, Power, and Action" and "This Ain't Another Statement. This is a Demand for Black Linguistic Justice." Likewise, the venerable WPA-L listserv has, for the first time in its long history, developed participation guidelines whose goal is, as Working Group member Iris Ruiz wrote to the listserv in a June 2020 email, "to move the list to a more equitable and inclusive climate."

Listening to BIPOC voices and those of people underrepresented in the academy describing the moves they must use to claim writing studies as discipline is an important goal for this book. The effects of racial politics, as Vershawn Ashanti Young and Keith Gilyard, among many others (Kynard, Perryman-Clark, Ore, and others), have at last become less elided, although work remains. The recent more focused (and belated) attention to underrepresented voices in the academy necessarily complicates questions about scholars claiming the discipline. In fact, these voices have yet to be heard as loudly as they should be, and this edited collection seeks to provide space for their important stories.

As Young demonstrates, the decision graduate students make to become writing studies scholars and claim writing studies as their own requires strategic thinking. While Young's rhetorical performance will be familiar to many scholars of color as a necessary tactic, writing studies scholars of color must not only perform but also engage with an unlikely discipline that, for some humanities scholars, lacks gravitas. This, too, shapes teacher-scholars' sense of how they may claim writing studies as their own. Chapters by Karen R. Tellez-Trujillo, Khadeidra Billingsley, Raymond D. Rosas, and Antonio Byrd detail their experiences as BIPOC writing studies scholars.

Precarity, a second troubling exigency, has increased in the last decade and it too surfaces in the chapters of this book. The term "precarity" here refers to the growing number of adjunct and non-tenure-track positions in the humanities and, especially, in writing studies. As Eileen Schell points out in her introduction to Seth Kahn, Willliam B. Lalicker, and Amy Lynch-Biniek's (2017) *Contingency, Exploitation, and Solidarity: Labor and Action in English Composition*, "the economic conditions of higher education have shifted over the past thirty plus years due to increasingly neoliberal, corporatized, and privatized models of higher education." Contingency and precarity are symptoms of this economic reality, pressurized by the COVID-19 pandemic economy. The number of advertised positions has been in a steady decline since 2010, and Jim Ridolfo's rhetmap (n.d.), which provides a list of available jobs each year, shows a precipitous drop during two years of pandemic and little sign of recovery.

Meanwhile, writing programs and departments dependent on contingent labor attempt to locate ethical practices while facing the shrinkage of humanities disciplines in general. Two chapters from this collection foreground this complicated space: "Embracing Failure: A Newly Independent Department's Attempts at Writing Its Own Script" by Brooks, Dadas, Fields and Restaino and Ersheid, Koenigsburg, McVeigh, Pearson, and Kahn's co-authored "Claiming and Being Claimed by Writing Studies: Negotiating Identities for Creative Writers Teaching Composition."

This essay collection sought narratives across three categories: (1) researched narratives about arriving to the discipline, especially those stories that detail how, why, and under which institutional, cultural, or social constraints professionals made writing studies their scholarly and pedagogical foci; (2) historiographic or ethnographic studies of writing studies program-building either within or outside an English department; and (3) theorized examinations of the strategies necessary to claiming the discipline of writing studies. The eleven chapters of this book represent a strong sampling of the categories.

The book is divided into three sections. They are Kairos and Opportunity, Negotiations and Resilience, and Allegiance and Identification. These sections emerged organically from accepted contributors' drafts; like focuses, experiences, and stories were quickly evident across essays. The section titles do not identify strategies for claiming writing studies. Instead, they identify the conditions the writers identified as essential to their own experiences. However, this is not to say that these themes or conditions exist only for the titular section. Rather, they represent strands that run through every essay; they help structure the stories, but their colors dominate more clearly. For example, "negotiation" can be mapped across all eleven essays, whether between faculty member and institution, institution and state government, student and teacher, or writer and sense of self. The twenty-one voices in this book (I include Cedillo, who wrote the final words, and myself in this count) recognize each of these conditions as integral to our experiences as teacher-scholars seeking to claim writing studies as our own.

Section I: Kairos and Opportunity

The first section represents authors for whom the experience of claiming the discipline has been structured by opportune time and place. The right time and place are important to every story about claiming writing studies and always require commitment to a writing studies identity, of course. But the authors in this section call on them forcefully and, to some extent, fiercely.

Lauren Bowen and Laurie Pinkert provide important historical context in their "Political, Personal, and Pedagogical Imperatives: Tactical Disciplinarity among Early Members of Writing Studies." Focusing on the strategic moves needed to claim the discipline, Bowen and Pinkert describe results from their interviews of twenty-seven retired writing studies scholars. Their analysis provides a first look at how the fraught or nonexistent recognition of disciplinary identity necessitated careful attention to kairos and to space for opportunity so that these early writing studies scholars could situate themselves in a just-coming-into-being discipline. Put differently, these are, as Bowen and Pinkert note, "origin stories" whose narrative arcs can help elucidate later origin stories.

Suellynn Duffey's standpoint as a scholar working in the field since the 1970s takes a long view on claiming the discipline that is, nevertheless, built on what she calls the "embodied and relational." Duffey argues for the importance of stories in order to identify key disciplinary birth moments but also asks that scholars refuse the notion that the current moment exists without connection to all the moments before. Kairos and opportunity permitted Duffey's explorations, and her commitment to rhetoric gave her impetus to seek out scholars outside of the literature department, as did many scholars, before writing studies emerged as discipline.

Duffey notes the significance to the CUNY open admissions policy and the important "Students' Right to Their Own Language" declaration, among other initiatives, noting, however, the shortcomings that failed to empower racially minoritized students as hoped. Above all, Duffey's chapter demonstrates how commitment to student learning has always been a pivotal strategy to claiming a place for writing studies.

The ellipsis or invisibility of writing studies has not, unfortunately, dissipated as Tara Wood's "Strategizing Disciplinarity, Disciplinary Strategies" chapter demonstrates. Writing from the position of a newly hired writing program administrator (WPA) for a struggling first-year composition program, Wood argues for what Laura Micciche has called "slow agency" (quoted in Wood, this volume). While state and university mandates constrict Wood's curriculum revisions, she also finds moments of opportunity to enact first, slow changes. Committed to antiracist assessment, Wood engages in deep reflection in order to locate kairotic moments where, at the least, conversations about antiracist practices can occur. For WPAs, Wood finds, claiming writing studies as discipline requires attentiveness to "the when and where."

The last chapter in this section, "Embracing Failure: A Newly Independent Department's Attempts at Writing Its Own Script," by co-authors Ron Brooks and Caroline Dadas with assistance from Laura Fields and Jessica Restaino, takes Wood's argument about the complications of working within state and university constraints a step further as they document the labor and birth of a writing studies department. An unfriendly English department and supportive upper administration delivered an opportunity, kairotic in its meshing of right time, right place, and right people. However, as Ron Brooks, the founding department head learned, the desire to be utterly collaborative in form ran into challenges much akin to what Wood experienced. Their essay draws on queer methodologies because this approach allows for analysis of the development and circulation of norms as well as imaginative tactics embracing messiness and allowing failure. The chapter breaks ground in the newness of the experience it describes but also demonstrates how well queer methodologies can enrich the story and give rhetors a way to name their experiences.

These four chapters instantiate the attentiveness necessary to writing studies scholars as they claim their discipline. Moments for action may slip by easily. Identifying and acting on these kairos-infused opportunities unquestionably marks a key strategy for claiming the discipline.

Section II: Negotiations and Resilience

The four essays in this section trace the many negotiations and state of resilience necessary to claiming the discipline. Kairos and opportunity remain discernible in this section. In "Claiming and Being Claimed by Writing Studies: Negotiating Identities for Creative Writers Teaching Composition," Alison Ersheid, Lisa Koenigsburg, Maureen McVeigh, Nancy Pearson, and

Seth Kahn write from varied positionalities: The first four are creative writers whose working lives are embedded within writing studies, unlike Kahn who is a tenured writing studies scholar. A focus on teaching writing, elemental to the fields of creative writing and writing studies, links these co-authors as well as a commitment to student learning. These five, like so many, negotiate identities as union activists, poets, and writers as well as teacher-scholars. The chapter identifies these negotiations to the backdrop of material labor and the possibility of tenure for that labor, dangled and withdrawn for years. While the story ends happily, the tension between claiming and being claimed by resonates throughout.

As in Ersheid et al.'s essay, Cynthia Johnson acknowledges the complications of labor as a writing studies scholar in her chapter "From Pell Grants to Tenure Track: Precarity and Labor as a Disciplinary Pathway." Johnson notes that her privilege as a white person clashes with being queer and from a disadvantaged class background. Her whiteness helped make it possible for her to mask the potentially "unacceptable" aspects of herself, yet her desire to mask these and gain professional status led her to accept extraordinarily demanding terms of labor and compensation. As she argues, financial precarity and affective labor may inflect the lives of graduate students across disciplines. However, within the field of writing studies, while the unspoken commitment to heavy material labor remains, effective and persuasive arguments deconstructing notions of professionalism, destandardizing language, have been emerging and these, Johnson says, helped her claim her place as a writing studies scholar. These commitments have enabled her to negotiate what feels impossibly tangled and, finally, to speak.

The scholars in this section see resilience as integral to their efforts to claim identities as writing studies teacher-scholars. In "Finding Resilience in Writing Studies at the United States-Mexico Border," Karen Tellez-Trujillo, like Johnson, finds resilience through the work of the writing studies scholars she has studied. Living among borders has costs, especially given the dominating and aggressive presence of white male scholars in her graduate student experience, and so the voices of feminist scholars in particular have given her space to make her claim. Now a new assistant professor, Tellez-Trujillo sees language as both freeing and imprisoning and, having written her way through to resilience, looks for the same markers of struggle and achievement in her students' writing.

Section III: *Allegiance and Identification*

This section includes four scholars whose life experiences have inspired a keen sense of what allegiances they require from the discipline. While contributors in sections I and II name explicit claiming strategies, section III contributors foreground their embodiment as underrepresented members of academia and commitment to social justice and equity through which they locate the language they evolve and story they create. For these contributors, strategizing identification resonates as crucial affective labor. Likewise, for them, locating opportunity and finding resilience are necessary conditions to their stories.

In "Being the Only One: The Embodiment and Labor of Tokenism," Khadeidra Billingsley describes her experiences as a Black woman and graduate student at a predominantly white school. Her embodiment in white eyes has enforced the status of token, and Billingsley movingly explains the weight, burden, and cost of that tokenism. Her essay responds to Black feminist scholars such as Kynard, Perryman-Clark, Robinson, and others whose refusal to be silent and commitment to their identities inspires. For Billingsley, telling her story will not only provide strategies—and, perhaps, comfort—to young scholars facing similar burdens but also represents her commitment to the discipline and to making a path amid "complexity and confusion." Telling her own story emerges as a key tactic in the face of the burden heaved onto her back.

Billingsley's allegiance to Black women who have come before and will come after her despite the everyday violences she experiences as "the Only One," echoes in Raymond Rosas's "Literacy and Disciplinarity: Vignettes of Struggle and Identification." For Rosas the violences arise first through language. These are inscribed through his embodiment as Chicano/Puerto Rican and experiences as classed and raced and as a survivor of the US opiate epidemic. Rosas locates his allegiance to writing studies through his commitment to equity in linguistic and literacy practices and sees the discipline as a site where such initiatives are truly possible. Rosas's reasons to choose writing studies as discipline push him to his chief strategy: he intends to write himself into the discipline and allow himself the power of hope.

In "Cognitive Dissonance," Alison Wells Zepeda explores the tensions arising from her embodiment, beliefs, and values. She believes in the power of language; like Rosas, Zepeda privileges how language has informed and shaped her identity and the questions she has asked of herself and of the discipline. Yet she also values affect, material feminisms, and the interactions between thing and environment. These complicated relationships as well as Zepeda's continual returns to language as primary suggest that she, like Billingsley and Rosas sees writing into and around the discipline as her once and future strategy.

Antonio Byrd's chapter, "Writing into Inclusion from the Margins," echoes the calls from Billingsley, Rosas, and Zepeda for ethical languaging. The murder and violence occurring in Ferguson, Missouri, honed Byrd's awareness of the margins of the discipline and underscored the necessity of traversing his racial positionality and the whiteness embedded in the discipline. Byrd investigates his own story and considers how his literacies were first informed by white writers whose scholarship dominated his early educational experiences. Not until his doctoral studies and Ferguson did Byrd realize how linguistic racism had shaped his literacies and, like the other authors in this section, he came to read and write himself as Black man and as a writing studies scholar into the discipline.

Christina V. Cedillo closes the book by amplifying the importance of story and highlighting how embodiment shapes and is shaped through writing studies. Cedillo underscores—as does the accumulative power of the book's chapters—the significance of story as pivotal strategy in claiming writing studies.

While all the chapter contributions engage with the question of pivotal strategies, their approaches, points of tension, and subjects differ. Likewise, methodologies differ depending on what questions the contributors chose; these range from qualitative mixed method (Bowen and Pinkert) to autoethnography (Byrd, Rosas, Billingsley, Tellez-Trujillo among others) to institutional ethnography (Brooks, Dadas, et al., Kahn et al.) to personal reflection and autobiography (Duffey). Each section opens with a brief interlude or introduction foregrounding the themes and conditions integral to the section.

These scholars' critical awareness of their embodiment and identity inspire their writing. They do not, however, only turn inward as they write. They are keenly aware of the politics and cultural expectations of the moment: they struggle to value their own voices, knowing they, like the scholars who came before them, will always need to adapt, always need to stand up, and always be in the moment of claiming the discipline.

References

Armstrong, Cherryl, and Sheryl J. Fontaine. 1989. "The Power of Naming: Names That Create and Define the Discipline." *Writing Program Administration* 13 (1–2): 5–14.

Gilyard, Keith. 1999. "African American Contributions to Composition Studies." *College Composition and Communication* 50 (4): 626–644.

Heilker, Paul, and Peter Vandenberg, eds. 2015. *Keywords in Writing Studies*. Logan: Utah State University Press.

Kahn, Seth, William B. Lalicker, and Amy Lynch-Biniek, eds. 2017. *Contingency, Exploitation, and Solidarity: Labor and Action in English Composition*. Fort Collins, CO: The WAC Clearinghouse. https://doi.org/10.37514/PER-B.2017.0858.

Lauer, Claire. 2012. "What's in a Name? The Anatomy of Defining New/Multi/Modal/Digital/Media Texts." *Kairos* 17 (1). http://kairos.technorhetoric.net/17.1/inventio/lauer/index.html.

Olson, Gary A. 2002. *Rhetoric and Composition as Intellectual Work*. Carbondale: Southern Illinois University Press.

Phelps, Louise Wetherbee, and John M. Ackerman. 2010. "Making the Case for Disciplinarity in Rhetoric, Composition, and Writing Studies: The Visibility Project." *College Composition and Communication* 62 (1): 180–215. http://www.jstor.org/stable/27917890.

"Rhet Map: Mapping Rhetoric and Composition." n.d. Rhet Map. Accessed March 1, 2021. http://www.rhetmap.org/.

Rodrigue, Tanya. 2013. "A Portrait of a Scholar . . . In Progress: An Interview with Louise Wetherbee Phelps." *Composition Forum* 27 (Spring). http://compositionforum.org/issue/27/louise-wetherbee-phelps-interview.php.

SECTION I

Kairos and Opportunity

INTERLUDE ONE

Kairos and Opportunity

Bowen and Pinkert; Duffey; Wood; and Brooks, Dadas, Field, and Restaino describe program-building experiences claiming writing studies at the institutional level. In effect, these essays zoom out to offer snapshots of individual (Duffey; Wood) and collective (Bowen and Pinkert; Brooks, Dadas, Field, and Restaino) experiences effecting change at the university level. Time, both linear and dynamic, frames the essays. Bowen and Pinkert analyze the stories of retired teacher-scholars while Duffey recounts her own story across time. Wood considers both the contexts of her university and the slow, steady, stubborn moves necessary to her work as a WPA with expertise in writing studies. Likewise, Brooks, Dadas, Field, and Restaino reflect on the challenges and rewards of establishing a stand-alone writing studies department. For these contributors, success hinged on the availability of internal (resourcefulness, openness to change, resilience, belief in the value of the discipline) and external (friendly administrators, supportive colleagues and allies, university commitments to writing) resources.

Seeking and finding the opportune or kairotic moments to effect change was essential to these stories—a dominant thread running through and informing how and when movement could be possible. Kairos is dynamic and its shifting offers, evident in these essays, underscore how politics shape and constrain. Yet, also, what Duffey calls "the long view" is evident here. That is,

https://doi.org/10.7330/9781646426331.p001

these stories look back before and as they look forward. The opportunity to strategize emerges across these stories not only because of kairos and opportunity but because the actors readied themselves and seized the day.

But, in fact, the outlines of these stories are familiar. They give us angles and form to what may feel both cloudy and formless. They matter, especially because they reveal the conditions through which shifts become more than nascent, claims to a discipline concretize, and program-building becomes possible.

1
Political, Personal, and Pedagogical Imperatives

Tactical Disciplinarity among Early Members of Writing Studies

LAUREN MARSHALL BOWEN AND LAURIE A. PINKERT

Introduction

This chapter explores the cultural habitus that shaped some individuals' decisions to claim a disciplinary identity when writing studies lacked many of the markers of disciplinary legitimacy. Efforts to trace the origins and history of writing studies have often followed the cultural and rhetorical conditions through which the discipline emerged, identifying the development of graduate programs (Chapman and Tate 1987), scholarly journals (Connors 1984; Goggin 2000; Hesse 2019), and CIP codes (Phelps and Ackerman 2010) as important evidence of disciplinary growth. Only occasionally (Elliot and Horning 2020; Roen, Brown, and Enos 1999) has scholarly discourse made visible the stories of the individuals who worked to "[earn] the respect of others" and "make the case for disciplinarity" (Phelps and Ackerman 2010, 180–181) as the field developed and professionalized. Here, we report on interviews of twenty-seven retired members of rhetoric, composition, and writing studies (RCWS) who were moved to claim their disciplinary identities at a time when the discipline was actively forming.

In reporting these stories, we hope to represent them to contemporary generations of writing studies scholars and teachers in order to promote

https://doi.org/10.7330/9781646426331.c001

cross-generational engagement with the similarities and differences that our disciplinary positionings afford. To that end, we do not present these analyses of interviews as blueprints for careers in writing studies today, nor do we anticipate that readers will necessarily "recognize that they are like the scholars and teachers" represented here (Roen, Brown, and Enos 1999, xvii). Instead, we examine how claiming one's place in the field during the early days of writing studies was a tactical response to cultural and historical contexts. Similarly, Suellynn Duffey's chapter in this volume discusses in further detail how her own "disciplinary identities intersect with institutional, economic, and political circumstances." Our intention is to highlight how social and cultural conditions prompted decisions to identify as a writing studies scholar, teacher, and/or administrator at a time when disciplinary infrastructure was still in its nascent form. In this way, we hope to document a historical moment by which current and upcoming members of the discipline might consider the kairotic elements of their own disciplinary identifications—as, in this volume, Antonio Byrd notes when recalling that a friend's introduction to a WPA became the first time he'd heard "rhetoric and composition" named, and as Maureen McVeigh describes when teaching first-year writing became a gateway through which she "fully claimed [writing studies]" for herself. This historical work can offer a sense of united purpose, meaning, and historical grounding, and aid cross-generational dialogue that acknowledges the relationships between the conditions of disciplinary identification and the tensions, disruptions, and fissures experienced among various generations of writing studies scholars and teachers.

As a follow-up to a related study on the experience of retirement (Bowen and Pinkert 2020), we focus in this chapter on the experiences of claiming initial membership in the discipline during an early, formative period for both the field and our interviewees. We ask: How do these retired scholars and teachers articulate the conditions and rhetorical strategies for affiliating with writing studies and, in doing so, how do they conceive of the field itself, both as it was and as it is now?

To address this question, we look at responses to semistructured interviews of twenty-seven self-identified retired writing studies teachers and scholars. In order to examine how interviewees rhetorically construct their early claims to disciplinary identity as rhetoric, composition, and/or writing studies scholars or teachers, we analyzed participants' responses to two of our ten standard interview questions:

- How were you first introduced to the field of rhetoric, composition, and/or writing studies?
- Has your relationship to the discipline changed over time, either because you have changed or because the discipline has changed—or both?

Interview participants were recruited from among respondents to the Survey of Intellectual Labor and the Academic Lifecycle, which we circulated on professional listservs, at CCCC gatherings, on social media, and on CCCC caucus mailing lists (Pinkert and Bowen 2019). After completing the survey, respondents were invited to indicate interest in participating in a further interview. Thirty volunteers self-identified as retired, and twenty-seven of those were able to complete the interview. Interviewees represent a range of retirement ages (68 to 85), disciplinary specializations, and institution types and sizes across the contiguous United States. The group represents several disciplinary cohorts: those who completed dissertations and founded programs in the 1970s; the cohorts that followed closely in the 1970s and early 1980s; and a subgroup of participants who entered the field later in their post-doctoral careers, in the late 1980s and 1990s. Two-thirds of participants self-identified as female, and one-third male. All participants identified as white, non-Latinx/Hispanic/Spanish—a lack of racial diversity that generally reflects the homogeneity of cohorts entering the professoriate in the latter decades of the twentieth century.

In reporting on these interviews, we do not suggest that all of our participants used *writing studies* to name and, in turn, claim their disciplinary home. In our original call for participants, we invited those who self-identified as members of "Rhetoric, Composition, and Writing Studies"—a deliberately capacious label. Our participants did not call this broad term directly into question unless they felt truly at the margins of the field, such as those positioning themselves primarily in technical communication or history of rhetoric. Regardless, we find their theoretical, methodological, and/or epistemological allegiances aligned with the contemporary work of writing studies. In this way, our study contributes to the scholarly conversation on disciplinarity by focusing on early associations with a nascent field as a tactical response to a particular cultural habitus, to which our participants responded agentively in seizing opportune moments to shape their careers and develop disciplinary identification in writing studies.

The retired writing studies scholars we interviewed all achieved success in the field, according to traditional measures, with many having achieved tenure, published prolifically, and held prominent leadership roles. At the same time, many characterized their disciplinary origin narratives as accidents. While we did not prompt our interviewees to comment on this particular trope, we call attention to the happy accident narrative because it highlights the impact of the emerging disciplinary status of writing studies—as a "pre-discipline"—on the career trajectories of these scholars. Due to the field's fledgling status, few of our interviewees would have been able to adopt a truly strategic approach to claiming a place in the field.

Although this volume is dedicated to the articulation of pivotal strategies, this chapter emphasizes somewhat more *tactical* dimensions of individuals' claims to disciplinarity. Michel de Certeau's (1984) well-known distinctions between practices with strategic and tactical aims is a useful frame here. In de Certeau's formulation, a *strategy* marks the "calculated" practice of an established body of power, with "a place that can be delimited as its own" (35–36). In contrast, "a tactic is a calculated action determined by the absence of a proper locus," emerging from "the space of the other." Glossed as the "art of the weak," a tactical approach "must accept the chance offerings of the moment, and seize on the wing the possibilities that offer themselves at any given moment" (37). As interviewee Erika Lindemann described, "There was no there, there" when most of our interviewees were launching their academic careers in what would cohere as writing studies. Thus, in claiming disciplinary identity without (yet) a "proper locus," early writing studies scholars relied on more tactical practices, in which serendipity, luck, and happy accidents were not only common but also necessary to the growth and establishment of a new discipline. As readers consider these early scholars' experiences, we are hopeful that the tactical nature of *all* disciplinary identity development—then and now—gains visibility.

In the remainder of this chapter, we will examine the conditions and rhetorical framing of these twenty-seven scholars' disciplinary origin stories, which demonstrate the often-tactical nature of these academics claiming membership in a field at its beginnings. In these stories, we see individuals responding agentively to economic, cultural, and professional conditions in order to forge a new discipline and their places within it.

Literacy in Crisis: Political and Economic Conditions of the 1970s

For graduate students in English studies and adjacent fields, the 1970s was an exciting and volatile decade in which to enter academia in the US. For those who could pursue advanced degrees in writing studies, this work was both complicated and daring, as these scholars completed dissertations with a small number of rhetoric and composition mentors (often only one in their own institution), with most English department course offerings in literature (see Duffey, this volume), as there were no more than twenty PhD programs in rhetoric and/or composition, as compared with ninety-four cataloged as of this writing (Ridolfo 2023). In fact, simply counting PhD programs prior to 1980 is a methodological challenge, as there was no clear consensus about what constituted a PhD concentration in rhetoric and composition (Chapman and Tate 1987; Covino, Johnson, and Feehan 1980; Lauer 1980). In describing the early efforts to complete composition dissertations prior to the proliferation of doctoral programs, Janice Lauer (1995) noted that "these pre-disciplinary acts seem pretty gutsy because they boldly positioned composition theory as appropriate subject matter for graduate study and, more importantly, they began to enlarge the web, introducing others into a community that was in its earliest phase of disciplinary formation" (281).

Further, in a period marked by economic downturn and political controversy, institutions of higher education became pivotal sites of national debate. Most notably, growing public perceptions of a literacy crisis, heralded by the publication of *Newsweek*'s incendiary "Why Johnny Can't Write" (Sheils 1975), was largely fueled by the "great decline" of mean SAT scores since the 1960s (Fleming 2011, 179). While the credibility of this claim of decline is dubious, fears of literacy's downward slide nonetheless motivated institutions' strategic investments in the study and teaching of developmental writing courses; these institutional responses created new sites of practice in which up-and-coming scholars could develop disciplinary identities.

Disciplinary identification is influenced not only by scholarly literature and professional discussions that comprise their sense of disciplinary connection but also by the institutional "enterprises" (Clark 1998) in which they work on a day-to-day basis. Disciplines and institutions converge in a "space of action" (Bauer et al. 1999)—most commonly for writing studies experts, in an English department or independent writing program. Our interviewees found themselves in the midst of an opportune moment to claim expertise in writing studies as teachers, tutors, administrators, and researchers, which would position them not to provide an answer to a badly formed question about literacy but to coalesce a discipline around better, more nuanced, and more culturally savvy knowledge of language and writing pedagogy.[1] The economic climate of the period created significant challenges for higher education, many of which served to amplify a sense of crisis in college writing instruction. Facing economic problems including inflation and a predicted shortage of traditional students, colleges and universities (supported by civil-rights-era calls for affirmative action) began opening their doors to broader populations, increasing the presence of formerly underrepresented minorities (Henze, Selzer, and Sharer 2007, 14).

In this economic and political climate, composition classrooms became a primary locus of controversy from which English departments could "work out their own difficulties" and respond to emerging pedagogical demands (Henze, Selzer, and Sharer 2007, 28). While some departments responded by doubling down on current-traditional pedagogies, others became breeding grounds for composition scholarship, most prominently yielding the process movement. The field's professional organizations also responded to literacy crises, with the most notable example being the CCCC resolution "Students' Right to Their Own Language" (Committee on CCCC Language Statement 1974). For some emerging members of the field, claiming and enacting writing expertise was a tactic for achieving dual aims: social justice and disciplinary identity.

Rebecca Mlynarczyk, for example, took tactical advantage of the open admissions policies first adopted in 1970 by City College of New York (CUNY), a primary site for studies of basic writers (e.g., Shaughnessy 1977), to launch her career as a compositionist. Having relocated to New York in 1969 after earning her MA in literature, Mlynarczyk worked as a freelance editor but always had aims of being a teacher. As luck would have it, an acquaintance helped her to get a "foot in the door at CUNY" in 1974, when she was called to interview for a writing tutor job at Brooklyn College. As Mlynarczyk recalls, claiming the role of tutor-expert in the Brooklyn College preparatory program (what would now be known as a stretch program) provided her with access to the *real* experts on developmental writing: her students. She recalled,

> From the early 1970s to the 1980s [the program] was very successful.... Any student who wanted it could have access to a tutor or who was somebody like me with a Master's degree. So I feel that my mentors were these students. I get emotional about it. I just fell in love with these students and

working with them on their writing. They had so much to say, and they were my mentors. There were wonderful professors in the program, and I learned a lot from them, too, but honestly my mentors were these students and that's been the way, all the way along.

In teaching in a literacy-focused program at CUNY during open admissions, Mlynarczyk's work with developing writers became the subject of study for two doctoral dissertations in the field of composition, which "was being sort of invented" all around her. This tactical positioning put her at the center of revolutionary ideas about student writing and errors not only as a teacher and study participant but also as a researcher, completing her doctoral dissertation on the reading/writing journals of bilingual college students and co-authoring an early textbook from her early days at CUNY. Thus, she could use her role to contribute to an empirical and socially just response to the traditions of linguistic and pedagogical racism and classism in higher education.

Stephen Bernhardt was similarly well positioned to contribute to and benefit from the surge of political and economic interest in underrepresented college students and their literacies. While at the University of Michigan, Bernhardt witnessed "an emerging attention to Black students on campus," which eventually yielded a crucial windfall: grants from the Mellon and Ford foundations, which sponsored the establishment of the English Composition Board in 1978. As he reports in an earlier iteration of his disciplinary origin story, "The ECB was a multi-pronged response to the broad concerns about student literacies that were current in the culture at large"—a response that aimed to support writers and increase recognition of varieties of English and the expansion of rhetoric; this work was, for Bernhardt, "exhilarating" (2007, 43). As he told us in the interview, "the situation was right to make that [work in composition] happen" and, we add, to simultaneously spur disciplinary development for himself and for the field of rhetoric and composition. In this way, the institutional strategic interests in marginalized students presented researchers with a kairotic moment of disciplinary self-invention. In answering the call for new knowledge and expertise on postsecondary student literacies—with particular interest in marginalized student populations—researchers and teachers like Bernhardt claimed a direction not only for their *own* disciplinary trajectories but for the trajectory of the emerging field.

While economic and political circumstances created a sense of urgency for academic literacy support and spurred the emergence of a generation of composition-inclined faculty, some institutions became sites of politically and economically motivated backlash against the growing movement for increased—and reformed—composition instruction. In one instance, carefully documented by David Fleming (2011), the University of Wisconsin-Madison English Department's decision to abolish composition was in large part the result of continued tensions between graduate students and faculty. Following the Vietnam War protests of the 1960s, Wisconsin TAs formed the Teaching Assistants' Association (TAA), a union that advocated for the need to revise the sharply current-traditionalist first-year writing courses to make them relevant to students. The response from the English Department and the university—for a variety of reasons well-documented in Fleming's history—was to abolish the standard Freshman English course in 1969, a dramatic change that meant the UW campus "essentially had no composition requirement" until 1996 (2011, 195). Thus, for a generation of UW-Madison graduate students, the opportunity to consider rhetoric and composition from a pedagogical entry point was largely unavailable.

Personal Circumstances Yield Disciplinary Connections

Never wholly disconnected from the broader cultural conditions that galvanized a generation of scholars in the 1970s and 1980s, the circumstances of our interviewees' lives often had a great deal to do with their initial introduction to the field. As Raymond Rosas notes in his chapter of this volume, "Writing studies has an established history of connecting the dots between lived experience and constructions of scholarly selves." For many of the early scholars we spoke with, claiming disciplinary identity as a rhetoric and composition teacher and scholar was a tactical means of balancing career and family needs.

Edith Baker's narrative, as she shared it with us, highlighted the ways that personal commitments and responsibilities might forge disciplinary allegiances. In her decision to pursue a master's degree, Baker explained, "Money at that point was a deciding factor for me," and the funding she received from a Woodrow Wilson Fellowship allowed her to move out of state to Virginia, where she studied American literature. During her master's degree, she married and moved to a Navajo reservation with her spouse. When he was drafted into military service, she returned to Chicago to live with her parents, working full time at a community college, teaching myriad classes in composition, literature, and rhetoric. During this time, she resolved to pursue an advanced degree but was undecided between literature and composition. Having interviewed with both literature and rhetoric and composition faculty at the University of Arizona, her ultimate decision to pursue writing studies was primarily based on how the programs aligned with her personal circumstances. She got the impression that it would take "ten years to get through" a literature PhD, whereas the rhetoric and composition faculty with whom she interviewed were willing to give her some credit for her prior work and acknowledged the possibility of maintaining a family life while also being a student. She recalls, "The rhetoric and comp people [made the degree seem] much more doable, I guess, in terms of predicting that I could succeed without giving up a life. And at that point I was a single mother with a child to raise, so that was a factor in my choice, too, because I didn't want to totally disrupt [my child's] life."

Joyce Neff's recollections, too, highlighted the ways that her role as a parent prompted her to return to graduate school, though in a different way from Baker's. A self-described "lifelong learner," Neff took a secretarial job after completing her undergraduate degree in English, education, and library science. Then, personal circumstances prompted a need for change: "I had two children, [and] I thought, I need to do something with this brain." Thus, she began the long and arduous process of graduate study, eventually leading to her PhD in education from the University of Pennsylvania in 1991, where she wrote her dissertation on the development of a cross-disciplinary community college writing center, under the guidance of committee members Michelle Fine, James Kinneavy, Marvin Lazerson, and Susan Wells. Like Baker, Neff was motivated by her positioning as a parent, but her decision to pursue graduate school highlighted her personal need to continue her intellectual pursuits alongside her domestic ones.

The personal circumstances that motivated early writing studies experts to claim the discipline were often described and inextricably intertwined with the economic and political conditions described in the previous section. At the same time that college student populations were becoming increasingly diversified and public debates over language and literacy raged on, the prospect of a tenure-line academic career in English studies was bleak. In the wake of the US economic recession from 1973 to 1975, doctorates granted outstripped the number of tenure-line jobs available, and by 1978, only 40 percent of doctorate-holding jobseekers could secure a permanent, full-time faculty appointment (Henze, Selzer, and Sharer 2007, 19–20). It was in this climate that Debra Journet got her professional start, as she moved toward completing her PhD in literature in 1980. She recalled, "I did not think I was going to get a job because nobody was. This was in the late 1970s. I decided I would finish my doctorate because I just am committed to finishing things I start." While adjuncting for a first-year writing program at her husband's institution, a colleague advised Journet to consider teaching technical writing, as it "would give more teaching opportunities." Although Journet assumed it would be "boring," she took a seminar on teaching technical writing, which introduced her to rhetoric and opened the door for her to pursue work as an industry tech writer. When Journet ultimately went on the English faculty job market, she was invited to fifteen interviews at MLA: "Only one of them in British Modernism [her doctoral specialization], and the other fourteen were in tech writing." Her decision, like Baker's, was prompted not by a commitment to disciplinary allegiance but by a personal decision to pursue more and better opportunities for professional work. As with Cynthia Johnson (this volume), many interviewees traced their inroads to writing studies through socioeconomic pathways leading to "tangled positionality of precarity and privilege." As Johnson reminds us, the high-labor (and often low-pay) demands often make for a precarious labor force in writing studies. For these particular early members of the field, however, labor of composition seemed an accessible pathway into the professoriate.

The Pedagogical Imperative: Disciplinarity in/through Practice

While some of our participants found themselves on the writing studies path in the early days of their careers, most came to the field having already entered the professoriate. For many, writing studies responded to a workplace exigence: the need to teach student writers *well*. Here, then, the field was frequently claimed as a humanizing intellectual framework for the institutionally assigned labor of teaching and WPA work.

As it had been for Rebecca Mlynarczyk, the experience of working with basic writers was a revelation to Carol Haviland. Working with a writing program designed to support basic writers in a one-on-one setting, Haviland found that she loved such "innovative" teaching: "If teaching could be like this," she recalled thinking, "I might like to do it." Likewise, while Joe Martin was initially reluctant to claim the teaching of writing as his primary occupation, his temporary-turned-permanent position as a writing teacher and administrator redirected his plans for a PhD in literary studies. He explained:

> I came back to Cornell to resume graduate studies in literary studies, and this [composition teaching] job opened up. I thought it was going to be temporary. It was a way to support myself. And I found myself doing a different

kind of teaching, teaching for which I had not been prepared, and teaching for which I had do some very, very hard and fast learning to understand what the teaching of writing to the students really meant. I had taught writing, but I hadn't reflected on it, as most people in that era hadn't. . . . I've been spending my life just trying to catch up with how the field of composition has been developing.

Both Martin and Haviland exemplified the shared sense among our interviewees that the teaching of postsecondary writing to diverse populations requires a great deal of conscientiousness, experience, and knowledge. Recalling their time as novice writing teachers, interviewees described their early careers as shaped by a dawning sense that the teaching of writing marked both a space for social justice as well as new and rigorous intellectual territory. Given the close alignment between teaching practice and the scholarly work to inform (and transform) that practice, it is unsurprising that most of our interviewees saw an essential and total overlap between the teaching of writing and membership in rhetoric, composition, and writing studies. Nancy Mack, who said she was able to "sneak in" to writing studies by gathering dissertation data from her own teaching experiences, articulated the pedagogical imperative inherent in her own sense of disciplinarity: "There wasn't the separation between scholarship and teaching like there is for so many people. To me, the two were pretty much the same."

In addition to teaching, writing program administration was also cited as a common point of entry for writing studies, as early generations of English studies scholars suddenly became WPAs, often without any graduate education in writing studies. Joel Wingard recalled that, while he had always taught first-year writing, the early stages of his career had emphasized literature and journalism. However, following the departure of the composition program director, Wingard "felt like somebody needed to watch over" the program and so took it upon himself to become "a little more involved in first-year writing from the administrative standpoint." In the following decade, Wingard's university began to experiment with a new writing curriculum, which expanded the scope of first-year writing and Wingard's role as a WPA. This led him to becoming more involved in the field's professional organizations in the late 1990s, including the Council of Writing Program Administrators. Following the experimental writing curriculum, a decision was made to adopt a writing-across-the-curriculum approach. By then aware of writing studies, Wingard approached the dean and requested the hire of a WAC specialist. The dean, however, had a different idea: Wingard would direct the WAC program.

Knowing he would need to deepen his writing studies expertise in order to succeed in his new post, Wingard secured (for a time) funding for conference travel and books to immerse himself in a new field. From his encounters with the discipline, Wingard felt that he came away with an invaluable new ethic of teaching and deeply admired his writing studies mentors for their humanity as much as their teaching strategies.

Somewhat earlier in her own career, Jane Donawerth, who had been denied the opportunity to teach composition while a graduate student in literature at the University of Wisconsin-Madison, also developed writing studies expertise as a faculty member. As with Wingard, Donawerth took the learn-by-doing approach and volunteered as the WPA of the introductory writing program. Although she had expertise in the history of rhetoric and classical rhetoric, she was less prepared to lead the composition pedagogy seminar for graduate TAs. As she says, "I just self-taught myself" by inviting local composition experts to lead one class each: "They taught me the class and then I started doing all the readings." Having begun some self-education, Donawerth accepted a colleague's invitation to a regional composition conference. "Then," Donawerth says, "I was hooked."

Feeling at Home: Negotiating Disciplinary Boundaries

In sharing their origin stories, participants had a lot to say about their perceptions of the field's emerging sense of boundaries and unity. As previously described, participants who did not have formal graduate education in writing studies found a way to "sneak in" (in Nancy Mack's words) to the field through teaching and writing program administration. Many of these writing-studies-experts-in-the-making, however, faced disciplinary isolation at their institutions. For example, Debra Journet, having been hired to replace the only writing studies scholar in her department at the University of Louisville, described her early career as a tenure-line faculty member who "didn't have a clue about research in the field," having "only been trained in literature." This, naturally, felt like a "significant challenge" for Journet, but she began the process of "learning about rhetoric" and retraining herself by reading and teaching graduate courses in composition and cognition, in which she "always felt like [she] was basically one week ahead of them [the doctoral students]," or in history of rhetoric. Rather than finding mentors in her own department, Journet had to rely on her knowledge of science and her literary training, as well as the syllabi of colleagues at other institutions, in order to advance her expertise.

Despite their determination, persistence, and resourcefulness, many of the early scholars we interviewed expressed gratitude for writing studies conferences and collaboratives that reduced the sense of intellectual alienation. Jane Donawerth described her sense of rhetoric and composition as an open, welcoming discipline. Having formally trained in Renaissance literature, Donawerth—whose interest in women's studies has been, since the beginning, a cornerstone of her disciplinary identity—found that Renaissance literature conferences in the field "were not, in those days, comfortable places to be. There was a real sense of hierarchy and competition, and there was no Women's Studies." This issue extended to other spaces of action within Renaissance literature, including its disciplinary journals: "One of the early articles I sent out came back [and the reviewers said], 'This is a really interesting, clear article'—and it was on women—'but our readership would not be interested in this.'" However, after she began attending composition conferences following her appointment as a WPA, Donawerth encountered a completely different climate: "They [rhetoric and composition scholars] are still, of all the [conferences] I go to, the most liberal, politically savvy group in the particulars of academic politics, as well as in just more general approaches to how you see scholarship and how you relate. So it's just a wonderful—you know, I just felt more at home there."

The experience of RCWS as "home" was, for most interviewees, built on the sense that writing studies had developed an ethos of collaboration. Because early writing studies scholars had to introduce themselves and others into a field that was still finding its own center(s) and seeking institutional legitimacy, professionalization largely involved peer networks rather than institutionally sponsored training. Erika Lindemann recalled that, in her early participation at CCCC meetings, "if you were lucky," presenters would offer mimeographed bibliographies from their presentations, or "you'd 'sit around in a bar' and listen to other scholars 'pontificate about this, that, and the other'" and in that way discover new ideas and learn about new scholarship in the field. Along with her peers, Lindemann "put together our own kind of graduate course" by picking up on what other colleagues would willingly share. Echoing this very sentiment, Joyce Neff recalls her early conference participation, in which she realized, "here are the people who are doing what I'm doing, and they want to talk about it." Likewise, Joel Wingard says that the CWPA conference "had been my graduate school." For these early scholars—particularly those who came to the field post-graduate studies and found the teaching and administration of writing to demand new knowledge, access to writing studies conferences was essential to moving from a sense of solitary struggle to disciplinary belonging.

Prevalent among the descriptions of the collaborative learning community into which interviewees were initiated was a sense of intimacy and cross-disciplinary connectedness. Edith Baker described her first CCCC experiences as one of being part of a small but dynamic professional circle. She recalled, "I think my first [CCCC] was in 1983, and there would start to be a few people on the program, but you know, most people were still going to MLA rather than CCCC, so it was definitely an evolving time. There would be special interest groups that would meet, and you got to know a lot of the important people in the field." Lil Brannon, too, remembered feeling a sense of connection with leading members of the field at conferences, cramming six to a room when, as graduate students, she and her colleagues couldn't afford to do otherwise. "We would go to meetings . . . and that's where I met all the people like Maxine Hairston . . . all the main people at that time."

In descriptions of the field as welcoming, home-like, accessible, collaborative, and offering direct connections between novices and experts, we hear resonance with what Anne Ruggles Gere describes as "intimate practices." The sisterly ethos of turn-of-the-century women's clubs, on a national scale, enabled the clubwomen to "see themselves as part of a larger whole" (1997, 10). Likewise, interviewees describe similar intimate practices, which, to our view, reinscribes the discipline as "feminized" in the sense articulated by Janice Lauer: a field not only coded (and devalued) as the domain of women (Miller 1990) but also characterized by actions that are "cooperative, relational, interdependent and collaborative; releasing in others their unexplored resources and transformative power; viewing development as a web; caring for another's development; [and] suffused with desire and joy" (280). Emerging from our participants' accounts, we see a consistent claim that decorous membership in the field often involved a decentering of one's own merits and gains in favor of building a discipline as a coalition. Thus, while entrance to the field may have been tactical—or outright accidental—the emerging disciplinary spaces of action were essential to establishing a sense of unity.

Not all characterizations of membership in the field emphasized the qualities of collaboration, transformation, interconnectedness, and joy, however. Many of our participants acknowledged disciplinary tensions in English departments, echoing Maxine Hairston's well-known CCCC Chair's Address (1985): "We often find ourselves confronting the literature faculty who dominate so many departments, and we feel that we are fighting losing battles. . . .

You can mount an all-out assault and think you're making an impression, but when the smoke clears, nothing has changed" (273). Yet, the tensions within English departments often served to strengthen the dawning sense of unity within rhetoric and composition circles. As Lauer (1995) recalls in an earlier published account, "We traded stories about our adventures as local pariahs, and we shared ideas about writing" (281).

Recalling the tensions between literature and composition, Bernhardt explained, "So the troubled history of English departments and their traditional centering on literary, historical, critical studies, I found was always, I thought, an impediment. . . . Because people [in my department] didn't really understand how you could take a skill like writing and treat it as an academic discipline." Bernhardt's account highlights the felt sense among early writing studies faculty of being an outsider within their own departments and institutions. As Bernhardt further explains, "Being in comp/rhet has always been a marginalized or outsider status and often poorly understood, even within your own department." Lil Brannon also acknowledged the English studies rift, which—like Bernhardt—she interpreted as a result of being misunderstood and thus undervalued by her literature colleagues: "In the early days, the split was between lit and comp. The lit people, they saw nothing important about teaching writing. They just thought you were just kind of grammar queens or whatever. There was nothing. They looked down on the work because, of course, they were scholars, and we were not."

Others emphasized a sense of being wooed rather than driven away from literary studies—their first academic "home"—by the inherent challenges and rewards of writing pedagogy. Rebecca Mlynarczyk found that the pull to composition, which never diminished her love for literature, was primarily social and critical in nature. "I didn't have a strong enough desire to just go straight ahead [to a PhD in literature]. . . . But I really did feel drawn to this, working with these students and their writing." Also getting her professional start in literature, Nancy Mack found herself "intrigued" by the challenge of writing instruction, as she found teaching writing "to be so much more difficult to do" than teaching literature. Joel Wingard similarly noted the influence of teaching on his disciplinary realignments, which allowed him to continue to build a career around the act of writing as the central subject, but from what was, to him, a more compelling perspective: "I found [teaching writing] so much more rewarding than teaching literature. I also liked the idea of teaching how, not what: teaching people how to do something, not teaching them what something is."

But the literature/composition divide was not always experienced either as an institutional or intellectual split. Having been trained in literature or other fields but, over the course of a career, engaging with composition, some participants noted that the lack of formal training in writing studies was an advantage, which ultimately defined their disciplinary approaches to scholarship and teaching. Rich Haswell, who marked his "shift" to claiming composition when he became a WPA, felt that he was able to bring his prior disciplinary knowledge to bear as a compositionist. His commitments to history and empirical studies, he surmises, "really came directly out of the kind of training I've had in literature scholarship, and perhaps even to his earliest disciplinary training in archaeology," which he initially studied as an undergraduate. Haswell believed that the tensions between empiricism and cultural studies within composition matched the tensions between literature and composition—both of which he had to navigate over the course of his career. Ultimately, however, his sense of being a cross-disciplinary scholar, while steering him clear of participation in the disciplinary clash, left him feeling doubly alienated: "So to the very end, I was still always teaching half literature, half composition, so I felt marginalized because I just didn't compete in that angry rift. . . . My colleagues either hated the comp people or hated the literature people. And I felt sort of left out."

A Shifting Discipline

Given that most interviewees' entries into the field came at a formative moment in the discipline, we were curious to what extent they felt that the discipline they helped establish had evolved since then. To close out this chapter, we consider interviewees' responses to our question, "Has your relationship to the discipline changed over time, either because you have changed or because the discipline has changed—or both?" In reflecting on the respondents' sense of a shifting discipline, we hope to offer readers a chance to consider what values may continue to resonate, or even stand in tension with, the discipline that currently emerging cohorts of colleagues will endeavor to (re)create.

A vast majority of interviewees noted that the field has diversified its primary subjects and methods. Some welcomed this change. Janice Lauer, for instance, saw such diversification as a natural evolution: "The field grows, different people come in and emphasize different things. I've seen a lot of things and I wonder, 'Hmmmm, what's that about?' It doesn't bother me. This is the way the field has to work itself out." Nancy Mack, too, welcomed the change as the field grows and expands: "The field was smaller [then], and now there are so many offshoots in the field in different directions. . . . I still like the field." However, both Lauer and Mack described a sense that the field is progressing in a positive direction, which is not a universally shared opinion among their fellow senior scholars. Their sense was confirmed by the majority of our interviews; overwhelmingly, interviewees expressed a feeling that the field was moving away from the pursuits (the social-justice-leaning work of teaching postsecondary writing) that once marked its center of gravity, toward a usually (tactfully) unnamed but increasingly diverse set of concerns.

Lil Brannon noted that when she was first involved in the field, "it was very much [about] the teaching of writing, and solidly grounded in first-year writing, and so it seemed like if you were going to be involved in rhetoric and composition, part of your job was going to be to direct a writing program, or to be involved in some way through a writing center . . . and be involved in the conversation about teaching." As the field grew, however, she noticed that it "moved further away from issues of teaching and issues of the politics of first-year writing and . . . labor issues in first-year writing." While she agreed that there are still cohorts who are invested in such issues, she observed "a large group of people now who aren't as interested in that aspect of the work."

Mirroring this perspective, Erika Lindemann described that the division of the field "into so many subspecialties," which she saw as a product of the academic tenure and promotion system that rewards work on subjects, methods, and problems that are deemed new to the discipline. However, Lindemann felt that this drive to specialize has left the field at risk of losing its pedagogical "center":

> I think if we get so, so many sub-, sub-, sub-, sub-, subspecialties or begin thinking about too esoteric a field, then we've lost the common core that really holds us all together. And I worry about that possibility happening. I don't think it's happened [yet], but I see tendencies of it happening and it concerns me. . . . [Graduate students] are still spending 50% of their time for their entire professional lives in a classroom, so teaching has to matter. Your subject is important. Content matters. But in teaching writing, process is the content.

Others, like Rebecca Mlynarczyk, conceded the need to "make room" for new specialties in the discipline, the "core" of the field as it was identified in its formative years still matters: "I still feel that the core values that those of us of this age were developing in the '70s and '80s, I think are still foundational. . . . I hope the field doesn't lose that as a foundational value."

While many of our participants saw the emergence of subfields that were less connected to the perceived pedagogical center as potentially problematic, Stephen Bernhardt contrasted this perspective. He demonstrated the ways that the pedagogical center marginalized him: "I made my mark in technical and scientific and medical communication, and I spent a lot of time in hospitals and tech companies, and in pharmaceutical development environments. So that put me further on the outside." Having understood the field's initial focus as narrowly emphasizing school-based writing and pedagogy, Bernhardt, unlike many of his counterparts, described the rise of specializations as a positive effect of disciplinary development.

In relating their origin stories to us, and in offering their on-the-spot impressions of the field's trajectory over time, these senior scholars and teachers describe a field forged by the fires of economic and political upheaval, as well as those of personal circumstance. As the field matures, these scholars recognize the motivations of disciplinary work taking on new characteristics, which in turn makes new disciplinary identities and engagements possible. In their responses we observe an important trend that suggests a fundamental truth about the discipline of writing studies: while it may at times, for some, feel united, it has never *really* been stable. Our interviews and the chapters in this collection affirm that writing studies has always developed and continues to develop within "an ambiguous and highly flexible construct" (Mendenhall 2014, 12). While interviewees represent similar motives for claiming the discipline (e.g., writing pedagogy as a critical response to cultural and political challenges) and, for some, a shared disciplinary ethos (e.g., collaborative and practice-oriented), early scholars in the field were still scrabbling for disciplinary legitimacy emerging from spaces of action, largely writing classrooms and writing centers. Yet even then, the trend toward "fluid, varied and distinctive identities" was already true of writing studies faculty, perhaps as a product of a pedagogical and praxis-oriented "trans"-discipline, which in its formation did not hold a stable identity.

Perhaps just as prominently as the histories of tensions with literature, writing studies has a long history of grappling with questions of disciplinary diversification, particularly as it relates to the relationship between theory and pedagogical practice. This ongoing tension is most notably documented in perennial "theory wars" in our publications and conferences, in which the struggle over the field's "center" has continued (Dobrin 2011; Olson 2000). In recent years, however, the field has made inroads toward consolidation. As Kathleen Yancey (2018) proposes, a new "disciplinary turn" may now be upon us, as we see the field claiming and stabilizing new spaces of action—including writing studies majors at universities and the consolidation of disciplinary knowledge as threshold concepts. Even so, as evidenced by the many responses to such efforts, the constitution of a stable center is hardly settled, even if our field's generative locus has shifted (perhaps) from sites of practice to methodologies of knowledge-making.

The historical context of the narratives we discuss in this chapter animates broader conversations about the differences and similarities with the contemporary landscape, as new members position themselves within writing studies amid the present cultural, political, and institutional climate in higher education in an age of austerity. Still, it remains important that we view such individual histories as they are—retrospectives of lived disciplinarity—and engage with them to, as Cinthia Gannett and John Brereton describe, "face some of our own gaps and forgettings" as a discipline (2020, 149). There are some powerful master narratives that infuse (and gatekeep) the ways we—and especially our potential non-writing-studies colleagues—situate ourselves as members of *writing studies*.

Note

1. See Tara Wood's discussion in this volume of the ways first-year composition continues to create opportunities and challenges for disciplinary and interdisciplinary connections.

References

Bauer, Marianne, Berit Askling, Ference Marton, and Susan Gerard Marton. 1999. *Transforming Universities: Changing Patterns of Governance, Structure and Learning in Swedish Higher Education*. London: J. Kingsley.

Bernhardt, Stephen A. 2007. "Sidebar: Finding Composition." In *1977: A Cultural Moment in Composition*, edited by Brent Henze, Jack Selzer, and Wendy Sharer, 39–46. West Lafayette, IN: Parlor Press.

Bowen, Lauren Marshall, and Laurie A. Pinkert. 2020. "Identities Developed, Identities Denied: Examining the Disciplinary Activities and Disciplinary Positioning of Retirees in Rhetoric, Composition, and Writing Studies." *College Composition and Communication* 72 (2): 251–281.

Chapman, David W., and Gary Tate. 1987. "A Survey of Doctoral Programs in Rhetoric and Composition." *Rhetoric Review* 5 (2): 124–186.

Clark, Burton R. 1998. *Creating Entrepreneurial Universities: Organizational Pathways of Transformation*. Bingley, UK: Emerald.

Committee on CCCC Language Statement. 1974. "Students' Right to Their Own Language." *College Composition and Communication* 25 (3): 1–18.

Connors, Robert J. 1984. "Journals in Composition Studies." *College English* 46 (4): 348–365.

Covino, William A., Nan Johnson, and Michael Feehan. 1980. "Graduate Education in Rhetoric: Attitudes and Implications." *College English* 42 (4): 390–398.

de Certeau, Michel. 1984. *The Practice of Everyday Life*. Berkeley: University of California Press.

Dobrin, Sidney I. 2011. *Postcomposition*. Carbondale: Southern Illinois University Press.

Elliot, Norbert, and Alice Horning, eds. 2020. *Talking Back: Senior Scholars and Their Colleagues Deliberate the Past, Present, and Future of Writing Studies*. Logan: Utah State University Press.

Fleming, David. 2011. *From Form to Meaning: Freshman Composition and the Long Sixties, 1957–1974*. Pittsburgh, PA: University of Pittsburgh Press.

Gannett, Cinthia, and John C. Brereton. 2020. "Framing and Facing Histories of Rhetoric and Composition: Composition-Rhetoric in the Time of the Dartmouth Conference." *Talking Back: Senior Scholars and Their Colleagues Deliberate the Past, Present, and Future of Writing Studies*, edited by Norbert Elliot and Alice S. Horning, 139–158. Logan: Utah State University Press.

Gere, Anne Ruggles. 1997. *Intimate Practices: Literacy and Cultural Work in Women's Clubs, 1880–1920*. Urbana: University of Illinois Press.

Goggin, Maureen. 2000. *Authoring a Discipline: Scholarly Journals and the Post-World War II Emergence of Rhetoric and Composition*. Mahwah, NJ: Lawrence Erlbaum.

Hairston, Maxine. 1985. "Breaking Our Bonds and Reaffirming Our Connections." *College Composition and Communication* 36 (3): 272–282.

Henze, Brent, Jack Selzer, and Wendy Sharer, eds. 2007. *1977: A Cultural Moment in Composition*. West Lafayette, IN: Parlor Press.

Hesse, Douglas. 2019. "Journals in Composition, Thirty-Five Years After." *College English* 81 (4): 367–396.

Lauer, Janice M. 1980. "Doctoral Programs in Rhetoric." *Rhetoric Society Quarterly* 10 (4): 190–194.

Lauer, Janice M. 1995. "The Feminization of Rhetoric and Composition Studies?" *Rhetoric Review* 13 (2): 276–286.

Mendenhall, Annie S. 2014. "The Composition Specialist as Flexible Expert: Identity and Labor in the History of Composition." *College English* 77 (1): 11–31.

Miller, Susan. 1990. *Textual Carnivals: The Politics of Composition*. Carbondale: Southern Illinois University Press.

Olson, Gary A. 2000. "The Death of Composition as an Intellectual Discipline." *Composition Studies* 28 (2): 33–41.

Phelps, Louise Wetherbee, and John M. Ackerman. 2010. "Making the Case for Disciplinarity in Rhetoric, Composition, and Writing Studies: The Visibility Project." *College Composition and Communication* 62 (1): 180–215.

Pinkert, Laurie A., and Lauren Marshall Bowen. 2019. "A Brief Report on the Survey of Intellectual Labor and the Academic Lifecycle." Presented at the CCCC 2019 SIG for Senior, Late-Career, and Retired Professionals in Rhet/Comp/ Writing Studies, Pittsburgh, PA, March 15, 2019.

Ridolfo, Jim. 2023. "PhD Program Map." Rhet Map. http://www.rhetmap.org/doctoral/.
Roen, Duane H., Stuart C. Brown, and Theresa Enos, eds. 1999. *Living Rhetoric and Composition: Stories of the Discipline*. Mahwah, NJ: Routledge.
Shaughnessy, Mina P. 1977. *Errors and Expectations: A Guide for the Teacher of Basic Writing*. New York: Oxford University Press.
Sheils, Merrill. 1975. "Why Johnny Can't Write." *Newsweek*, December 8, 1975.
Yancey, Kathleen Blake. 2018. "Mapping the Turn to Disciplinarity: A Historical Analysis of Composition's Trajectory and Its Current Moment." In *Composition, Rhetoric, and Disciplinarity*, edited by Rita Malenczyk, Susan Miller-Cochran, Elizabeth Wardle, and Kathleen Blake Yancey, 15–35. Logan: Utah State University Press.

2
Through the Eyepiece (and Body) of a Long Past

SUELLYNN DUFFEY

Bushwhacking into Disciplinarity

I belong to the generation of composition and rhetoric practitioners and scholars who came of age before there was a (twentieth-century version of) rhetoric and composition, the generation[1] that Lauren Marshall Bowen and Laurie A. Pinkert's chapter in this book provides rich, collective information about and that I explore from an individual perspective. Several people of this generation have retired; some have died; and several have written their disciplinary origin stories, the common threads of which are that we hold doctoral degrees in literature and that we bushwhacked our way into professional lives—bushwhacked because no cleared paths existed into disciplinary territories where we might reap benefit. Just enough bits of pertinent scholarship had emerged in the late 1960s and early to mid-1970s to let us glimpse fabulously exciting inquiry that spoke to our teaching, allowed us to see our teaching as a site of scholarly endeavor, and imagine, if not locate, a modicum of community.[2] The revolutionary pedagogies that emerged to challenge current traditional ones were, however, scattered outposts in the wilderness territory we traveled. Spurred as we were—eager and sometimes desperate with the need to know—we bushwhacked our way into what became disciplinarity.

Yet, even though it is said that my generation invented a discipline, I doubt many of us were (initially) driven by that motive, or at least we didn't name our efforts in that way. As Bowen and Pinkert describe this period, "'Writing studies' lacked many of the markers of disciplinary legitimacy" (this volume), but it took time and scholarly discussions before we understood that part of what we sought was disciplinarity and institutional recognition. Our passion was writing and writing instruction. When institutional obstacles like funding for writing programs presented themselves, we learned to read the situational kairos and pivot with the resilience we had begun to develop, skills that continue to be necessary.

With my long life in this discipline as it was birthed and grew and in response to this book's question about how we claim our discipline, I focus on (re)claiming our history as we pivot toward the future and its challenges. I could cite any of the cliches about ignoring the past at our peril, but that is not my point. I'm especially concerned with how our discipline's histories are our inheritance, how we use that inheritance, and the ways in which individual, lived experience intersects with public knowledge and history. I'm not arguing about avoiding peril or paying homage to the past but instead for increasing depth, value, knowledge in our work—our scholarship, pedagogy, and community contributions. Kathleen Yancey's (2018) introduction to *Composition, Rhetoric, and Disciplinarity* (2018), about which I will say more later in this chapter, crystallized my thinking about how the past and present live within each other.

In keeping with this book's intentions, I begin with two stories that suggest ways in which the past is present in contemporary disciplinary concerns, an integration we do well to understand and use. The first story is about how disciplinary identities intersect with institutional, economic, and political circumstances and the second is about choosing doctoral study. I locate the first in a disciplinary discussion gaining national traction as independent writing programs began to appear.

What I know about the founding of a writing department then is a story less about a discipline asserting its identity because of its inherent value and much more about the kairotic situation in which the structural transformation occurred, one that involved available funds, administrative priorities, local institutional circumstances, and state government that supported its origin. Ron Brooks, Caroline Dadas, Laura Field, and Jessica Restaino's chapter in this book shows in more specific detail how creating a writing department is heavily entangled with institutional and political negotiations and demonstrates how kairotic moments are idiosyncratic to individual locations.

The second story also intersects with institutional instantiations of the discipline as well as how my experience of those intersections parallels aspects of graduate student experiences today.

Both stories then move this chapter into questions about how we value embodied and relational epistemologies, the costs we incur when we ignore them, and the benefits we might gain by working against narrow, rationalistic ways of knowing. Our disciplinary moment might well be one in which we can learn to value embodied knowing more fully as we also knowingly carry the past into the present.

Story One—Euphoria and Reality

When I joined Georgia Southern University's Writing and Linguistics Department, I vowed, in the silent and sometimes wordless way we often speak to ourselves, that I would never again work in a department of literature where the default topics of both informal hallway and other more substantive conversations were *literary* allusions, puns, conceptual frameworks, and critiques.[2] The default assumed that all of "us" spoke the same language, the disciplinary language of literary studies. In literature departments, I thought, composition and rhetoric programs and faculty might be accepted—often begrudgingly but sometimes with welcome.[3] Even so, the all-too-common default conceptual frameworks ignore—and *are disciplinarily ignorant of*—the alternate Others, their language(s), conversational interests, and expertise. The alternate Other that I was and am a part of. That others help constitute in ways different from how I do.[4] My new department marked for me not only one point in composition and rhetoric's coming of age, a hopeful sign of a promising disciplinary, departmental, and institutional future,[5] but also a milestone in my professional life. Hallway conversations would not cast me as Other, and the resilience I had developed to work in such conditions might take a rest.

I shared the euphoria of some of my new colleagues and others in the discipline at large; euphoria that arose from a sense of professional validation and comeuppance; a sense that we and the discipline were gaining visibility, equal rank, and status with literature departments; and institutional authority. My euphoria was not idiosyncratic, as Barry Maid (2002, 143) attests to: "Most narratives about the creation of independent writing units . . . are full of joy and hope," and I joined a newly created one with heightened expectations.

Such an idealistic view, while not completely inaccurate in its assumptions, ignores the kairotic forces that actually coalesced at Georgia Southern to support the assertion of disciplinary identity—or the appearance of it that a departmental identity proffered.[6] My euphoria blinded me to the complex context in which the department was created and the forces that initiated it, forces almost entirely unrelated to disciplinarity. Even though I concur with Louise Phelps and John Ackerman (2010, 181) that a "disciplinary identity is necessary for . . . [its academic work] to be taken seriously within the meritocracies of higher education and to help sustain the working identities of practitioners, scholars, teachers, and administrators across the United States," (2010, 181), I know that such a case, in order to effect certain kinds of change in academic institutions, needs more persuasive evidence than the value and identity of the discipline, those markers that inspire us in the field and drive many of us to desire independent writing studies departments.[7] It needs coins of the realm, and *Kairos* and exigence mint those coins and thus figure significantly into how such departments emerge, as much as or more so than disciplinary identity and desire.

As I studied my new departmental culture, its recent past (it was only a few years old at that time), and the forces that had enabled its creation, I understood that the coming of age aspect of my invented story was not only ahistorical but also as simplistic as any cultural myth we tell ourselves.[8] Instead of disciplinary comeuppance, bureaucratic forces at Georgia Southern and in the state's funding budget underpinned the department's creation, a department that was initially populated with faculty excised from the Department of Literature and Philosophy. Based on the history I could gather, it was neither faculty impulse nor scholarly, disciplinary assertion of identity that impelled the new department's creation. Far from it. It was the motives of higher administrators and the exigence of existing funding.[9] In addition, its early life was so conflictual that mediators were called in to help through its birth and early life (see Agnew and Dallas 2002). While my own negotiations at Georgia Southern engaged my professional identity, those that created the department were not about disciplinary identity and involved a complex ecology of forces. My euphoria was manufactured by forces very different from what I thought had engendered it.

I tell this story in our quest to understand what disciplinarity is, where it resides, how it plays out in the lives of scholars, practitioners, and other professionals who claim it, and how institutional frameworks intersect with disciplinary identity, intersections that have less to do with knowledge-making and other professional, pedagogical, and scholarly goals and negotiations than with bureaucratic and political ones. It behooves us to envision such broad and complex contexts when we imagine a disciplinary future.

Story Two—Seeking a Doctoral Program

My second story about contemporary graduate students calls forth surprisingly similar stresses in both their and my process of claiming the discipline through doctoral study—even though the stories and time periods differ substantially. In the 1970s, Susan P. Miller joined Ohio State's English Department and brought with her the fabulous new scholarship I mentioned earlier and whetted my appetite for more. But during my graduate career, the only composition-related courses available to me were a two-quarter linguistics course and a seminar on composition studies taught by Edward P. J. Corbett. In it, Corbett's (1971 [1965]) *Classical Rhetoric for the Modern Scholar* was a central text, one that had, I thought, very little pertinence to composition as I knew it. All the other departmental offerings were in literature.

Because the equivalent of special dispensation allowed me to take one course outside the English Department for my doctoral program, I took a course in rhetorical history that sealed my desire to leave the English Department and join the Rhetoric and Communications Department where I stayed for one year, learned rhetorical history from the classical to contemporary, spent a whole quarter studying (and fascinated by) speech act theory, and wrote a paper on women's language that my instructor thought publishable if I added some empirical evidence (which I was uncertain about how to do and so never published the paper). I also took a course in interviewing, one in statistics, and one in research design where I learned (imperfectly) to design empirical research projects and how to "operationalize" both dependent and independent variables.

Valuable as is the rhetorical and communication background I received in that department and the support I received for my efforts, its social science research methods taught me that I was much more at home in a humanities department. I returned to English, took my rhetoric and social science knowledges with me, and managed to focus half of my general exams on non-literature topics (psycholinguistics and comp/rhet). Support in the department allowed me to write a composition dissertation and include specialists in English education in my studies. Finding my way through a doctoral program felt more like stumbling than finding or even inventing a discipline as I worked through bureaucratic tangles and disciplinary hegemony.

Recently, I worked closely with two determined graduate students as they bushwhacked pathways toward doctoral study.[10] They had to handle uncertainty and pressures on their professional identity(ies) that I'm certain are not idiosyncratic. They had to negotiate constraints and opportunities in ways similar to mine. Did I really see myself as a social scientist? Or a humanist? Or maybe some combination of both that was in fact neither as they were disciplinarily defined?

For these students, deciding what programs to apply to was itself difficult. The costs were not only in the financial outlay that applications required. The students had to undertake deep reflection on their talents, abilities, and emerging professional identities, reflection that continually brought to light new insights and conclusions and could rebalance their scholarly direction. While they could name their current area of specialization, their interests would potentially change through doctoral study,[11] something I knew from lived experience.

These two students had to balance family considerations and, like the other three who sought doctoral study that year and many if not all in our program, they were first-generation graduate students and thus had to acquaint themselves, from scratch, with the structure of doctoral programs (course work requirements and elective opportunities, general/comprehensive exams, reading lists, dissertation defenses). They might have heard that a personal statement should identify scholars they wanted to work with, but they didn't know how to balance a single faculty's enticing scholarship with all the other aspects of doctoral study that would affect their learning, and so I helped them negotiate web-based and other information about programs they were considering because they needed help learning what questions to ask—of themselves, the information they were encountering, and the programs they found appealing. They also needed help finding information with which to project a professional future, like job placement statistics.

Doctoral programs are far from identical—in their course offerings, core programs, available faculty, design of qualifying exams, and preparation for the dissertation—but as the students I worked with read tantalizing course descriptions from their prospective schools, they were easily wooed. They needed help to read carefully for the underlying story(ies) a catalog of courses might tell. As we know, what a course promises—or ultimately delivers—may not be what the student expects, but more important is what a course catalog might indicate about programmatic opportunity and absences. For example, we found programs that allowed little flexibility in course work and others that allowed much.

We speculated about the effect of such differences on the students' doctoral knowledge-making. Some programs allow students to choose courses outside the department, a signal perhaps that the faculty valued cross-disciplinary work (important to one of the students). Alternately, it might signal that the program isn't well enough staffed to teach a full doctoral offering itself and thus its students might have very limited courses to choose from. A limited number of course offerings is not an insurmountable problem—dissertations can be written without the writer having had coursework in the area—but such conditions can set a harder task for the writers and possibly a minimized supportive framework.

In other words, the variety of doctoral programs—their number of course offerings in any given term, the number of faculty who can teach and the number of courses a department will allot to comp/rhet offerings (in relation to others it offers), the opportunity to have courses outside the department/discipline "count" toward the degree—all these are costs or gifts to weigh. Even in this time when doctoral programs have proliferated, some may not have the riches available to meet students' needs and desires; just as I did not when I was in graduate school. Now, more rhet/comp faculty and student peers are surely available to support such deliberative processes than were available to me, but all of these circumstances require the same degree of deliberation and educative, political, and professional uncertainty that I faced in choosing a program (with, of course, very different options).

The prevalence of programs is not a panacea, as I've also seen in one of our alums currently in a doctoral program where departmental politics and tensions trickled down into the student's decisions about what courses to take, advisors to choose, teaching assignments to seek, and so forth. And so, I wonder what differences our disciplinary growth creates for students at this level of professional life, entering and undertaking doctoral study. It seems to me that the similarities in decision-making that current students face entail the same or similar costs, tensions, and appeals at the level of individual experience (intellectual, emotional, professional), but the context of current disciplinarity and the choices it offers differ.

Reading History—Through the Body

I turn now to Kathleen Yancey's introduction to *Composition, Rhetoric, and Disciplinarity* (2018). Her work allows me to focus on thoughts about history, lived experience, and disciplinarity. Yancey's chapter maps composition through what she calls its historical episodes; "episode" being the term she chooses instead of "period" or "turn" or other similar descriptors in common parlance. Prominent during this early period, Yancey points out, was our field's linguistic focus when "teachers of composition, in the midst of teaching a group of students new to the academy, banded together to share knowledge about how to teach writing. . . . Their subject matter was language, their role teaching, their practice enhanced by borrowings from linguistics, itself a discipline eager to be applied" (2018, 17). Reading this historical narration, I *felt* truth in Yancey's words; truth that I had lived—experientially, bodily.

Yancey's words called up in me a long-forgotten image of a Reggie Rinderer, a colleague when I was in graduate school whose education-school knowledge of linguistics came with her to those of us teaching writing. I see her on Ohio State's campus where I did my doctoral study, amid a cluster of movable desks in a certain Bevis Hall room, where daylight poured from windows that reached the ceiling and created a distinct quality of light. This episode is also when Mina Shaughnessy visited my campus to advise us (Andrea Lunsford and my department) about creating a basic writing program, my memory of Shaughnessy including the advice she gave us as well as her elegance in the hallways we shared. My sense of that elegance is fully embodied, visual as well as kinesthetic, and it inflects my sense of Shaughnessy's advice. For me, then, the multisensory, embodied images of Shaughnessy and Rinderer thread through Yancey's linguistic episode.

My recollection of these two women and my mention of them here may suggest I'm engaging in sentimental, nostalgic time travel—remembering youthful experiences in ways that logo-centric and rationalist epistemologies easily dismiss. Instead, I want to emphasize my full-body experience of Yancey's (2018) linguistic episode in order to highlight what I see as one of our disciplinary challenges: How to add lived experience *more fully* into epistemological, scholarly, and pedagogical endeavors as we go forward. How to understand relational ways of knowing as, in part, an experience of the body. How to enrich epistemologies that guide all our endeavors with the texture, depth, and dimensions of understanding that lived experience adds to the scholarly.

My lived experience and embodied knowledge of Yancey's (2018) linguistic episode demonstrates a certain nuanced contribution to scholarly knowledge that lived experience affords. It's not only that I have a visual memory of Shaughnessy and Rinderer, a nostalgic overlay onto my knowledge of our discipline's linguistic episode; it's that when I read Yancey's discussion of the linguistic episode, I know it through my bodily as well as cognitive experience, and those dual channels add value to my knowledge.

Other voices in this volume and elsewhere—Indigenous, feminist, and those from minoritized communities—value this goal, and many of our publications reach toward it. Such views and practices privilege ways of knowing besides Western epistemological ones, ways of knowing that are not purely rationalist, that arise from and rest on embodied perceptions as well as cognitive ones. Embodied knowing includes the whole body and all its faculties in its epistemological work. Indeed, some would say that cognition itself is embodied, as I think it surely is. Relational knowledge positions the knower *in relation to* that which is in the process of becoming known instead of separate from the "object" of study. Indigenous world views and feminist rhetorical practices both align with these epistemological processes and are foundational to certain non-Western cultural practices and formations.

I'm not sure, however, that our epistemological theorizing has caught up with scholarly endeavors bringing minoritized voices into our conversations, particularly narrative's role in scholarship, teaching, and learning. Recent work on counterstory undertakes epistemological theorizing about the crucial importance of narratives and lived experience to rewrite and reclaim hegemonic theories, practices, and histories that exclude minoritized voices. But there's more epistemological work to do about lived experience and narrative, especially about how embodied and languaging epistemologies intersect with learning.

Open Admissions, Basic Writing Programs, and Antiracist Efforts

I now turn to another historical aspect of the linguistic episode as a very simple example of how lived experience and public history interweave and can together offer more, epistemologically, than either does alone. Concurrent with this episode were open admissions policies and the proliferation of basic writing courses and programs, one I directed and taught in. The birth of open admissions across the country and notably at the City University of New York (CUNY) offers, as well, another example of how *kairos* and exigence merge with other forces that lead toward disciplinarity. At CUNY, it was "increases in the availability of government aid for underprepared students, coupled with community demands for minority representation, [that] led CUNY to abandon its insistence on objective standards of college readiness and to implement a policy of access for all high school graduates" (Kimball 2021, 6).

By requiring only a high school diploma from applicants, open admissions policies signaled changes in the student populations that universities sought to educate. Shaughnessy's *Errors and Expectations* (1977) and the several biographies and autobiographies of people then at CUNY document the radical changes that open admissions brought—to students, classrooms, instruction, professional lives, scholarly research on writing, and institutional infrastructure.

At the time, such moves as CUNY's were seen as progressive, positive, and welcome in many quarters. As a young teacher in the seventies, I felt open admission policies to be revolutionary, partly because they pushed against an elitism implicit in academia that those who were first-generation students and/or from minority populations also felt—and still do, as we see in this volume. The era of open admissions corresponded with protests asking for more democratic governance on campuses, free speech initiatives, and the creation of Black and women's studies programs, among many other efforts like campus childcare facilities.

While basic writing programs seemed like progressive measures and officially sponsored and funded support programs for those who needed them, research began to reveal problems. For example, Eleanor Agnew and Margaret McLaughlin (2001) showed that exit exams in basic writing courses and/or programs disproportionately and repeatedly failed minoritized students and kept them in basic writing much longer than students from majority populations—and kept them there sometimes endlessly. By the 1990s, some of us were arguing for de-tracking the placement systems (and courses) that basic writing programs had engendered.[12]

As is well known, CUNY linked open admissions and racial integration, a history of the period stating that "the CUNY trustees viewed racial and academic integration as virtually synonymous" (Kimball 2021, 6).[13] This view parallels what had driven K–12 desegregation a decade or so earlier; a social, educational, and legal change effort with unintended and sometimes unfortunate consequences.[14] "CUNY's principal strategy for racial integration was to spread academically underprepared students throughout the university's 17 colleges, and to create a 'sizeable identifiable group' of the most severely underprepared students on each senior college campus" (Kimball 2021, 6). The term "underprepared," while accurate given the academy's white-centered language standards, can be read as troubling for that very reason. Implicit in assumptions around open admissions are racism and white supremacy (minoritized students will be "remedial" in their academic preparation rather than differently prepared).

And yet, viewed through the lens of exigence and kairos, we must read differently or at least inflect our reading with an awareness of the arguments necessary to secure funds. Nonetheless, while opening admissions was expected to increase the enrollment of minoritized groups—a desirable outcome—it also proffered evidence for the long-standing view that universities with open admissions had *lowered standards*, the conjoining of such standards and the enrollment of minoritized groups all too clearly, built on centuries of racial stereotypes.

In this disciplinary era, impulses toward social justice began to noticeably influence writing studies and its practitioners. "Students' Right to Their Own Language" was first adopted by CCCC in 1974 and serves as an example of the discipline's early focus on linguistic justice. In conflicted ways, CUNY, open admission policies, and well-intentioned basic writing programs can thus be seen as historical precedent for the field's current and much more robust antiracist concerns.

Many of us in composition classrooms, administering writing programs, and studying writing assessment and placement were committed to working against racism where and how we could, as the foci for our early institutional and professional negotiations suggest. Given kairotic situatedness, the foci were different, societal support then was different, the tools we had were different, and our strategies differed partly because scholarly and (white) cultural knowledge had not progressed as far as it has now, but it seems to me that the current antiracist imperative *alone* is insufficient to differentiate our contemporary disciplinary moment from earlier efforts toward racial justice in the field. This past inflects the present, certainly in my lived experience of it but also in the collective present. What seems different to me is that we (majority whites) are now situated within and surrounded by more complex understandings of white privilege, systemic racism, white body ignorance of minoritized cultures and histories, and how those understandings inflect our consciousnesses, goals, strategies. We negotiate racism with different tools, perhaps, and with different understandings of its methods and dispersal societally. But we were trying to do that during the discipline's linguistic episode, too; historical efforts that we do well to include an awareness of in our current efforts rather than dismiss.

This view leads me to call for a particular kind of historical work that conceptualizes current moments, efforts, plans, and/or desires within parallel and/or foundational historical moments as Elizabeth Kimball (2021) does in *Translingual Inheritance: Language Diversity in Early National Philadelphia*. There, she calls on John Trimbur and "traces the origins of [our very contemporary] translingual approaches in composition back to the advent of open admissions at the City University of New York in the late 1960s and 1970s, with Mina Shaughnessy's efforts to read student writing—marked by difference of all kinds, and not at all welcome to many within the university—on its own terms" (63). Kimball further traces this thread of initiatives and practices into the next several decades. She writes that Shaughnessy's "close reading practices of New Criticism, which she repurposed to subvert the literary establishment they were used to uphold . . . led to David Bartholomae's work in the 1980s, and to Min-Zhan Lu and Bruce Horner's work in the 1990s, all predicated on close reading of student texts" (Kimball 2021, 63). I find Kimball's (and Trimbur's) insights and arguments convincing—that a "familiar practice of close reading for finding, reading, and rereading the literate artifacts of past periods is exactly appropriate for retelling history in translingual terms" (63). In fact, Kimball's study of early Philadelphia's languaging milieu also rewrites the city's history away from a narrative that (counterfactually) supports Anglo origin and English-only stories into one that demonstrates how that narrative has blinded us to a long history of translingualism in Philadelphia and on our continent, a heritage that should inform our current efforts.

Reading Relationally through the Body

Let me return to Yancey's (2018) history. It lays out a progression of disciplinary episodes that seems to be linear, from linguistics to composing processes and then cultural theory, teaching, and disciplinary content, the last taking the majority of her chapter to explain, but the conceptual framework for her history is *organically accumulative rather than linear*. Indeed, her history is a disciplinary parallel to my reading experience of the history she writes. After she has differentiated the fourth "episode" from the previous three and each one of them from the others, she writes that in the "fourth episode, the field [is] still influenced by all the activities in the previous episodes" (2018, 17). Although it has sometimes seemed that any new episode has divorced itself from the previous ones—indeed, we scholars have talked about a new episode as we demonstrate its worth by denigrating the previous one's value—Yancey argues that the divorce image is wrong and, I say, misleading. Instead, each episode returns to and in some ways rests upon an earlier episode.

Even though the social turn seemed to dismiss process approaches, we know that process pedagogies still permeate writing classrooms, as they do mine. Writes David Fleming (2009, 40), "The teaching of writing in this country is nearly everywhere officially, if not actually, process-based." Chris Anson

(2014) also notes the deep integration of process pedagogies into classroom practice. Both Fleming and Anson analyze the shortsighted rejection of process as the field moved toward and through the social turn. Byron Hawk (2007) gives book-length attention to some ways in which the social turn failed in its attempt to reject expressive pedagogies.

More pointedly, I still assign process research in my contemporary teaching of the discipline. In classes on teaching writing, I include very old research on writing processes—that of Nancy Sommers, Sondra Perl, and Muriel Harris—because since the process era, scholarship hasn't provided such concrete images of how people actually write.[15] Graduate students who read this work in my classes still experience a revelation and newfound freedom about their own writing processes and gain, as well, insight about how to teach writing to others.[16] But, under some conceptions of scholarly value (the impulse to "remain current"), my practice can be seen as (at least) outdated, ignoring the best and newest that's been learned, and it can thus be construed as counterproductive to teach. The new-is-best model is possibly more applicable to research in scientific fields that build knowledge cumulatively in ways that humanistic study doesn't exactly replicate, but what is newest in our field creates an aura nonetheless that it is best, most valuable, deserving of the most attention, an impulse that has its limits.

Let me point to Yancey (2018) again—that each new episode in our discipline's life incorporates the previous ones. If so, instead of saying we should either ignore the old or throw out the new, we need to understand how they are interwoven. My lived experiences of them cannot be replicated in younger bodies, of course, but it calls attention to a richness, a depth that the imperative to keep contemporary might obscure.

Yet, my bodily and cognitive perceptions show me that I need each prior episode of the discipline and its knowledge as I participate in the current one or else the current one is diminished—it feels thin, precariously perched on a tiny island of knowledge instead of drawing from a scattered archipelago connected either by land pathways at low tide or bridges or ferries. I want—I need—connections (neural pathways in my body?) to more of the archipelago than one island, and my disciplinary epistemology deepens if the archipelago is instead a greater land *mass*—a peninsula at least, or even a continent.

My disciplinary knowledge is shaped like the archipelago I seek and its formation a spiralic progression, a progression that may break one side of the spiraling double helix, loop itself backward and outward, open up and reach sideways and crossways to draw in more threads as it moves onward in a more complex design of multiple spirals that have spun into different shapes. The complex spiral marks a pattern my knowledge of the field enacts. Like the winds of a tornado, this epistemological spiral pulls matter from the ground (or an earlier episode) into new spaces and adds material from other spaces into its vortex as it moves upward, forward, outward, and even backward.

Although I am not a tornado, and tornadic imagery is more violent and disruptive than the phenomena I describe, my reading experience mimics the tornado's movement, a process Yancey (2018) also aims to describe more placidly. My disciplinary experience creates for me—and for all of us as well—a problem located in scholarship, practice, professionalism, and embodied knowing, or if not a problem, then at least an issue to explore, a point that I believe underlies this book's genesis. If our field/discipline is at a pivot point, what does the field/discipline rest on, what does it move toward, what does it carry forward, and, especially, how does it sweep its past into the swirling vortex—with destruction, abandonment, accumulation, integration, or reconstruction?

The time humans have spent on this earth (among many other things) shapes their sense of the discipline, its scope and focus, its reach outward beyond its boundaries, and its depth. Its sense of what counts as knowledge. My sense of the discipline has been developing since the late twentieth century, and its spiraling knowledge carries pieces of disciplinary past into the present, sometimes haphazardly and sometimes deliberately.

So, as I end, I draw this conclusion: We need to enact Yancey's (2018) historical observation and my tornadic activity any time we apprise ourselves of where we are now. We need to look backward and spiral like tornadoes as we move forward. We need to fully enact this practice. Hopefully more peacefully than constructing crisis-ridden weather. I'm not saying that the past is the same as now nor that the future is merely the past in different clothing. But because I value embodied knowledge, lived experience, and relational epistemologies, and because I see the current discipline through the eyepiece and body of a long past, I'm committed to strategies of living and creating disciplinarity that are as rich as they can be. Because the past lives in the present, as Kimball's (2021) book shows, I see a broad role for history in our contemporary scholarship. I want our scholarship to emulate the connectedness Kimball shows, integrate the past into the present, and build the robust knowledge this integration enables. I see the past as supporting contemporary efforts and purposes rather than something old that should be dismissed and forgotten. I want our contemporary scholarship and discipline to embrace the archipelago of knowledge I feel I need.

Notes

1. Bowen and Pinkert in this volume show that this generation is actually comprised of several generational subsets.
2. My new department chair drew my attention to this conversational phenomenon.
3. I am certainly overgeneralizing on the basis of English departments I have known. While it is possible to assign the hegemony of literature that was departmentally true thirty years ago, we might be inclined to think the days of token rhet/comp appointments in literature departments are over. Yet, I suspect not, especially if we consider the full range of institutional types where English departments exist. In my current one, for example, a colleague had, as of 2017, never heard of English studies or the diverse model of disciplinary interests it suggests. In other words, while literature and comp/rhet might be well integrated in many English departments across the continent, change happens in scattershot ways, not in uniform, linear, integrated progressions, and some literature departments have yet to admit their first rhet/comp scholar, as evidenced by an advertised position in 2019. As Debra Frank Dew and Alice Horning's (2007) book, *Untenured Faculty as Writing Program Administrators: Institutional Practices and Politics*, suggests, the departmental place of rhet/comp scholars varies significantly, according to a complex array of conditions.
4. Identity categories like race and gender, constructed though they are, create circumstances that differ exponentially from my life as a comp/rhet person in a literature department, but in both cases, we Others have to develop strategies to position ourselves in ways that we find healthy, feasible, and ethical, and we are constrained in doing so.
5. See an example of the seduction independent writing departments hold in "Wagering Tenure by Signing on with Independent Writing Programs" (Crow 2002).
6. Other origin stories, other examples of independent writing departments, show the troubling lives they can have. See *A Field of Dreams* (O'Neill, Crow, and Burton 2002) as well as later publications.
7. The contributors to *Composition, Rhetoric, and Disciplinarity* (Yancey 2018) offer fuller analyses of processes that establish independent writing programs and the multiple disciplinary, institutional, and professional functions that departmental independence serve.
8. An excellent example of a recently deconstructed myth is the first Thanksgiving story, a historically inaccurate as well as damaging story that schoolchildren have learned since roughly the mid-nineteenth century as described by David J. Silverman (2019) in *This Land is Their Land: The Wampanoag Indians, Plymouth Colony, and the Troubled History of Thanksgiving*. In it, Indians help Pilgrims survive near starvation and celebrate together peacefully at the end of harvest time. We carry on the tradition of cooperation and giving thanks. This myth seriously ignores historical fact and erodes the Indians' complex diplomatic and political reasons for helping the Pilgrims. Silverman traces these reasons for what the myth describes as simple support. He also traces the myth's nineteenth-century origins; the cultural, economic, political, and nationalistic needs it served; and the erasure of Indigenous history that it still supports.
9. Because the book *A Field of Dreams* (O'Neill, Crow, and Burton 2002) was not yet published when I joined the department, my research was archival and somewhat informal. It uncovered these facts, explained in brief in "Internal Friction in a New Independent Department of Writing: And What the External Conflict Resolution Consultants Recommended" (Agnew and Dallas 2002). Certain higher administrators believed the Department of Literature and Philosophy had grown too large (though I was never able to ascertain exactly why that was considered problematic) and state funds were available for certain administrative costs. Hence the creation of a department of those who had been teaching lower-level writing courses, most of them without terminal degrees. Those with terminal degrees held MFAs or PhDs in literature, and some of them were not eager to leave a

literature department. In other words, the department I entered was not, at its formation, what I imagined it to be—a collection of composition and rhetoric scholars (even if it was a collection of composition practitioners). By the time I joined, several with doctorates in composition and rhetoric had been hired, but they were a minority in the newly extended family, a "blended" one with conflictual backgrounds, goals, and vision, as Agnew and Dallas describe.

10. My institution offers only the MA degree, so all of the doctoral programs they considered were elsewhere.
11. Part of my role, I realized, was to nudge them toward programs that offered at least a modicum of flexibility and could sponsor what I knew of their interests.
12. See, for example, my article, "Mapping the Terrain of Tracks and Stream" (Duffey 1996).
13. Kimball's (2021) "The History of Open Admissions and Remedial Education at the City University of New York" is a New York City government document.
14. Stereotypes about student preparedness surely rested on well-intentioned rationales as well; for example, the financial impoverishment of many schools for African American children and those in areas with a weak tax base. With hindsight, however, we know that just as school integration allowed racial stereotypes to be expressed in new guises, it also brought about unintended consequences, like the loss of jobs for African American teachers. Argued as a means of equalizing and thus improving education for African Americans, it also marked "good" education as "white" as Catherine Prendergast's (2003) book, *Literacy and Racial Justice: The Politics of Learning after Brown v. Board of Education*, lays bare.
15. Restricted as Sommer's, Perl's, and Emig's research base and the demographics of their subjects are, theirs (and others') still provide insight into the everyday processes many writers experience. Coupled with caveats about diverse writing processes and political, economic, and other such constraints on writing, their research provides important insight for writers and learners.
16. While writing process research of the late 1970s and 1980s did not address issues of diversity, it nonetheless provides insight that counters much current school teaching about how to write, beliefs and teachings lodged deeply in students' understanding unless they have been exposed to composition studies indirectly through their K–12 teachers or directly via an undergraduate major in writing studies. Almost none of the students (and high school teachers) I teach have any such background. I suspect that the backgrounds of my students are not idiosyncratic.

References

Agnew, Eleanor, and Phyllis Surrency Dallas. 2002. "Internal Friction in a New Independent Department of Writing: And What the External Conflict Resolution Consultants Recommended." In *A Field of Dreams: Independent Writing Programs and the Future of Composition Studies*, edited by Peggy O'Neill, Angela Crow, and Larry W. Burton, 38–48. Logan: Utah State University Press.

Agnew, Eleanor, and Margaret McLaughlin. 2001. "Those Crazy Gates and How They Swing: Tracking the System that Tracks African-American Students." In *Mainstreaming Basic Writers: Politics and Pedagogies of Access*, 2nd ed., edited by Gerri McKenny (and associate editor Sallyanne H. Fitzgerald), 85–100. Mahwah, NJ: Lawrence Erlbaum.

Anson, Chris. 2014. "Process Pedagogy and Its Legacy." 2014. In *A Guide to Composition Pedagogies*, 2nd ed., edited by Gary Tate, Amy Rupiper Taggart, Kurt Schnick, and H. Brooke Hessler, 212–230. New York: Oxford University Press.

Corbett, Edward P. J. 1971 [1965]. *Classical Rhetoric for the Modern Scholar*. 2nd ed. New York: Oxford University Press.
Crow, Angela. 2002. "Wagering Tenure by Signing on with Independent Writing Programs." In *A Field of Dreams: Independent Writing Programs and the Future of Composition Studies*, edited by Peggy O'Neill, Angela Crow, and Larry W. Burton, 213–229. Logan: Utah State University Press.
Dew, Debra Frank, and Alice Horning, eds. 2007. *Untenured Faculty as Writing Program Administrators: Institutional Practices and Politics*. West Lafayette, IN: Parlor Press.
Duffey, Suellynn. 1996. "Mapping the Terrain of Tracks and Streams." *College Composition and Communication* 47 (1): 103–107.
Fleming, David. 2009. "Rhetorical Revival or Process Revolution? Revisiting the Emergence of Composition-Rhetoric as a Discipline." In *Renewing Rhetoric's Relation to Composition: Essays in Honor of Theresa Jarnigan Enos*, edited by Shane Borrowman, Stuart C. Brown, Thomas P. Miller, and (assistant ed.) Sarah Perrault, 25–52. New York: Routledge.
Hawk, Byron. 2007. *A Counter-History of Composition: Toward Methodologies of Complexity*. Pittsburgh, PA: University of Pittsburgh Press.
Kimball, Elizabeth. 2021. "The History of Open Admissions and Remedial Education at the City University of New York." In *Translingual Inheritance: Language Diversity in Early National Philadelphia*. Pittsburgh, PA: University of Pittsburgh Press.
Maid, Barry, M. 2002. "Creating Two Departments of Writing: One Past and One Future." In *A Field of Dreams: Independent Writing Programs and the Future of Composition Studies*, edited by Peggy O'Neill, Angela Crow, and Larry W. Burton, 130–152. Logan: Utah State University Press.
O'Neill, Peggy, Angela Crow, and Larry W. Burton, eds. 2002. *A Field of Dreams: Independent Writing Programs and the Future of Composition Studies*. Logan: Utah State University Press.
Phelps, Louise Wetherbee, and John M. Ackerman. 2010. "Making the Case for Disciplinarity in Rhetoric, Composition, and Writing Studies: The Visibility Project." *College Composition and Communication* 62: 180–215.
Prendergast, Catherine. 2003. *Literacy and Racial Justice: The Politics of Learning after Brown v. Board of Education*. Carbondale: Southern Illinois Press.
Shaughnessy, Mina P. 1977. *Errors and Expectations: A Guide for the Teacher of Basic Writing*. New York: Oxford University Press.
Silverman, David J. 2019. *This Land Is Their Land: The Wampanoag Indians, Plymouth Colony, and the Troubled History of Thanksgiving*. New York: Bloomsbury Publishing.
Yancey, Kathleen Blake. 2018. "Mapping the Turn to Disciplinarity: A Historical Analysis of Composition's Trajectory and Its Current Moment." In *Composition, Rhetoric, and Disciplinarity*, edited by Rita Malenczyk, Susan Miller-Cochran, Elizabeth Wardle, and Kathleen Blake Yancey, 15–35. Logan: Utah State University Press.

3
Strategizing Disciplinarity, Disciplinary Strategies

TARA WOOD

At my current institution, a regional, public university where I serve as writing program administrator, the second course in our first-year sequence is titled *College Research Paper*. The title connotes an emphasis on research-based writing, an emphasis that is very common across first-year writing programs in the United States (see Isaacs 2018). On one hand, this commonplace may seem an obvious and expected choice for the shape of writing instruction at any given college, particularly in the first year. We need to prepare students for the rigors of writing the college research paper. On the other hand, I am compelled by a comment made by Elizabeth Wardle in a Macmillan Learning (2014) promotional video for the second edition of her co-authored textbook *Writing about Writing*:

> We've all known and felt as composition teachers for a long time that something just isn't right; something isn't working.... What we're asked to do in first-year composition is not a reasonable thing. We're being asked to inoculate students so that they write well for all settings forever. And anyone who's ever taught writing or tried to write knows that that's just not a thing that can happen.... It's a relief to learn that this feeling that you have that what you've been doing in your composition class isn't working is a reasonable thing to feel, that research supports that, and there is something else you could do.

I have experienced that relief firsthand as a composition instructor, but feeling it as a composition instructor and reconciling it as a writing program administrator are two different beasts entirely. There is a risk to saying out loud that what you are doing in first-year composition (FYC) is not working. There is a risk to admitting our task—as it has historically and commonly been defined—is ultimately and fundamentally unachievable. Such admission could be just the right line item on an administrator's budget: one less course, a reduction of adjunct labor costs, less time for students to graduation, more room for electives, more room in the general education program for other, more pressing, required curricula. The additional challenge to such an admission is to convince all campus stakeholders that the new idea is a better one. If I want to change our course title, *College Research Paper*, I will have to present my rationale with enough rhetorical finesse so as to avoid the inevitable and predictable response: "How can you say that research writing doesn't matter in college?!" Moreover, how can I simultaneously claim my discipline of writing studies while also working to build a Writing across the Curriculum/Writing in the Disciplines (WAC/WID) program on my campus? In any given day, I may find myself in one meeting explaining the importance of a writing studies (discipline-based) approach to FYC and in a later meeting drawing on service-based rhetoric in order to make arguments for the benefit of a WAC/WID program. What rhetorical tactics can WPAs strategically employ when claiming writing studies as a discipline? If interdisciplinarity is an epistemological cornerstone in the discipline of writing studies, how can such complexities be conveyed rhetorically to stakeholders across campus and beyond? Put another way, this chapter seeks to explore the tensions between disciplinarity and interdisciplinarity as they emerge in the rhetorical situations of WPA work. I share my own experiences of reshaping FYC curricula via a disciplinary approach while also working to advance an interdisciplinary WAC/WID program on my campus. This two-pronged approach has yielded some success in terms of claiming disciplinarity within my program, but the ability to leverage the value of writing studies disciplinarity within WAC/WID continues to be a challenge.

Within the last ten years, rhetoric/composition, or writing studies (WS), has been working to identify, map, and articulate our disciplinary identity. Phelps and Ackerman (2010) reported on the success of the Visibility Project, an effort to "gain recognition for the disciplinary study of writing by focusing on the ways that fields of instruction and research are identified, coded, and represented" (184). Arguing that "external validation matters," they reported on two measures of success in this effort for recognition:

1. Designation as an "emerging field" in the National Research Council's taxonomy of research disciplines.
2. Assignment of a "code series to rhetoric and composition/writing studies in the federal Classification of Instructional Programs (CIP)." (180)

These external measures have run parallel to numerous internal articulations of our discipline. Efforts to define our disciplinarity have yielded exploration of threshold concepts (Adler-Kassner and Wardle 2015, 2019), a focus on transfer-based pedagogies (Yancey, Robertson, and Taczak 2014), articulation of key words for writing studies (Heilker and Vandenberg 2015), and arguments over the nature of the writing studies major (Balzhiser and McLeod 2010; Scott and Meloncon 2017). The recent publication *Composition, Rhetoric, and Disciplinarity*, edited by four leading scholars in writing studies, opens by framing the exigence as our field's current positioning on the "cusp of disciplinarity" (Malencyzk et al. 2018, 3). In this collection, Yancey names the phenomenon I have briefly outlined above, suggesting that composition has undergone a "disciplinary turn" (16). She identifies four trends influencing this "turn":

1. A renewed research agenda, including research into and theory about transfer of writing knowledge and practice;
2. the development of projects consolidating what the field has established as knowledge;
3. the continuing development of the major in Rhetoric and Composition; and
4. the changing location of Writing Studies within institutional structures. (17)

Put succinctly, Yancey (2018) points to our disciplinary-focused research and the way the content of our discipline becomes institutionalized. While I agree with Yancey's characterization of the disciplinary turn, reconciling our internal characterizations with our localized institutional experiences is a significant challenge, a tension perhaps felt most keenly by writing program administrators. We are continually defending our disciplinary expertise, and yet claiming disciplinarity may obfuscate some of our greatest strengths (such as the interdisciplinarity foundational to many WAC/WID programs). As Adam Banks stated in his 2015 CCCC Chair's Address, we might imagine "composition programs and departments operating more like interdisciplinary centers than as programs and departments" and that "it is precisely because we do discourse, in all its messiness, that we have the chance to be this kind of hub for intellectual work and for justice work, on campus and off." His invocation of interdisciplinarity echoes what many have argued about the strength of our discipline (e.g., Bazerman 2011); that is to say, the generative possibilities of writing studies emanate from our lack of drawn boundaries. It is precisely because we cross disciplinary boundaries that some of our most provocative scholarship has emerged. But the undisciplined nature of our discipline has consequences when working with stakeholders on our campuses. When we leave the confines of our departments and of our classes, when we have to engage with colleagues and administrators across campus, the argument that our discipline is interdisciplinary is a rhetorical tactic with potential consequence.

For those scholars working within this disciplinary turn, addressing our interdisciplinarity is fraught. For example, in Adler-Kassner and Wardle's (2019) *(Re)Considering What We Know*, their opening chapter addresses critiques launched at the first edition (*Naming What We Know* [2015], hereafter *NWWK*). The first critique they seek to address is "Threshold concept theory focuses on boundedness between disciplines rather than connections and interdisciplinarity" (Adler-Kassner and Wardle 2019, 17). While *NWWK* sought to articulate our discipline's knowledge, a key critical response to such efforts bemoaned the absence of interdisciplinarity. Adler-Kassner and Wardle are quick to point out the institutional ramifications of avoiding disciplinary boundaries; they write, "While fields like writing studies have been informed by a number of other disciplines, there are beliefs, orientations, and research findings from our field that set it apart from other fields. Not recognizing this expertise . . . has many implications. Some of these are associated with institutional decisions" (18). They go on to list a number of examples, including issues of faculty lines, budgets, resources, and working with faculty across disciplines (18–19).

The three discourses I invoke in these opening paragraphs—a shared sense of the "problem" (or impossibility) of FYC expectations, the disciplinary turn, and my challenge of sleeping at night while administrating a course titled *College Research Paper*—all serve as the discursive backdrop for the exploration I aim to undertake in this chapter. I would like to explore the vicissitudes of disciplinarity and interdisciplinarity, specifically interrogating to what end each might be leveraged strategically in the service of writing program administration. That effort follows the distinction Bowen and Pinkert make in their opening chapter of this volume between tactic and strategy, as I hope to illuminate my own "calculated" attempts to situate my discipline with my own context. I will take up my own university's historic and current institutionalization (and/or lack thereof) of writing, writing instruction, writing studies scholars, and writing program administration.

The Value of FYC: Disciplinarity for the Discipline's Sake?

In my current role as WPA at a regional public university,[1] we have recently undertaken significant revision efforts to improve our curriculum, staffing, professional development, and assessment practices. This process has been a strategic effort to positively reposition our writing program in relation to the rest of the university. In the past ten to fifteen years at my institution, an intense amount of distrust and dissatisfaction with English writing offerings (general education writing courses) resulted in a chipping away of English Department control over writing instruction. While nailing down the direct or singular cause for such historic distrust of FYC and its effectiveness for students is difficult if not impossible, I suspect it can be connected to what Robertson and Taczak refer to as our "un-disciplinarity." In their piece, "Disciplinarity and First-Year Composition," they argue that "we are not a discipline but an 'un-discipline' . . . a field without consistent content in the introductory course representing our area of study, without consensus about research-based curricular approaches to FYC, and often without expertise behind the delivery of our FYC courses" (2018, 186). They connect the "undisciplined" nature of our field to external devaluation by stakeholders.

At my university, this resonates significantly. During my first year as WPA at this university, I met with any and every stakeholder that would give me time and perspective and quickly identified two commonplace perceptions of FYC: (1) no consistency of delivery or content equals problems for students and (2) poor teacher qualifications/expertise equals poor quality of instruction and accountability. While there are truths and untruths embedded in these threads, using each as a framework for understanding the perceived value of FYC in my role as administrator is useful. Each of these discursive threads—one on the content and one on the delivery—reflect common questions in most scholarship focused on articulating rhetoric and composition as a discipline. To wit, the trends I heard from administrators, staff, and faculty outside my discipline were the same topics we are currently discussing and debating internally within our own discipline: What is FYC and who should teach it? My initial reaction to this onslaught of anecdotal historic distrust was somewhat annoyed by the predictability of it all, and yet, as I reflected, I found myself wondering why I felt annoyed by stakeholders identifying that which we ourselves are working to pin down. How can we ask others to trust what we deliver and by whom if we ourselves are unsure? Many people I spoke to conveyed dismay at our heavy reliance on adjunct labor and TAs. However, as Robertson and Taczak (2018) point out, "given the broad range of expertise and backgrounds from which FYC programs draw instructors, and the challenges faced by an often transient and underpaid workforce at no fault of their own, it's understandable that individual FYC courses are delivered according to what an instructor's experience dictates or for which the reality of an instructor's workload allows" (192). Upon my arrival to this institution, the bulk of FYC courses were taught by contingent faculty, almost none of whom held credentials in the discipline of rhetoric and composition. In fact, most were homegrown, alumni of the graduate program in literature within the English Department.

In sharing this consideration of my own experiences as a WPA, I work to align myself with other WPA scholars who promote the value of storytelling, reflection, and solidarity building (see, e.g., Vidali 2015; Micciche 2007; Rose 2004). In *The Writing Program Administrator as Researcher* Shirley Rose and Irwin Weiser (1999) remark that "effective WPAs reflect before acting, but they also reflect upon the actions they take" (ix). My challenge as a WPA was to rebrand the writing program and regain trust from the institution, all while attempting to internally and externally claim writing studies as my discipline. I would like to share several key rhetorical tactics that I believe have garnered me some success in this effort, and I will also reflect on the tensions inherent to such tactics.

Wrangling WAC/WID: The Vicissitudes of Disciplinarity and Interdisciplinarity

As I mentioned previously, the distrust in writing instruction at my institution resulted in a chipping away of English Department control over general education writing courses. Disciplines and departments around campus created their own version of required composition courses at the first-year level with no oversight or input from English. These courses were put through general education curriculum committee and received little to no pushback, and in some cases may have even been encouraged or applauded for their initiative. In some ways, this resulted in a type of WAC-lite, but with no sense of WAC pedagogy, no direction or central anchoring, and no sense of the criterion for what might constitute writing-intensive (WI) courses. These discipline-based courses were first-year general education writing courses born of mistrust—a mistrust connected to the devaluation of FYC content and delivery as well as a complete lack of awareness or acknowledgment of writing studies as a discipline in its own right.

In order to claim writing studies as a discipline in my role as WPA, I took several deliberate actions. I will describe each in turn and work to demonstrate how each draws on rhetorics of disciplinarity and interdisciplinarity in an effort to promote the writing program and effective writing instruction at my institution. To begin, I started with the content of our curriculum. Before my arrival as WPA, the department chair in English was the historic lead in the composition program. My institution requires six credit hours in written communication (as articulated by the state department of higher education). However, those six credit hours can take the form of either a two-course sequence in the lower-division or a two-course sequence across lower- and upper-division courses. Historically, my institution offered a two-course sequence and no upper-division offerings in (general education) written communication. The first course, *College Composition*, is administered exclusively via the English Department, and the second course is offered by four different departments across four separate colleges. No central administrator presides over the entirety of the program. Any curricular transformation I might create would apply only to the English courses over which I have (some) jurisdiction. I opted for a writing-about-writing (WaW) pedagogical approach (Wardle and Downs 2016) along with some aspects of the teaching-for-transfer approach (Yancey, Robertson, and Taczak 2014). The second course would have a stronger emphasis on academic research, as is quite common across first-year writing nationally (see Isaacs 2018). I designed a common syllabus and common assignment sequence and worked with our Instructional Design team to also create a course-management system (CMS, Canvas) shell available to all instructors. I received considerable praise across the institution from various stakeholders (advisors, student success, upper administrators) for these efforts to standardize the curriculum. In all of my conversations, however, it was far easier to convince people of the benefits of standardization than it ever was to convince them of standardizing the curriculum to a writing studies approach. That is to say, they loved seeing standardization but did not quite care or clearly understand what it meant to standardize the curriculum around the content of my discipline. I considered renaming the courses but ultimately decided on what Laura Micciche (2011) calls "slow agency," or "focus[ing] not on tasks or deliverables but on pace, the rate at which activity progresses" (2011, 75).

While WPA work is often associated with "big agency" or "actions that intend structural results and effects" (Micciche 2011, 73), slow agency focuses on pacing. In order to effectively claim writing studies as my program's content (and by extension claim writing studies as my discipline), I recognized I would need to focus on the long game and work to strategically (and rhetorically) pace myself and the change I pushed for on campus. Standardizing the curriculum would garner me some amount of trust across campus. Stakeholders acknowledged the implementation of a common syllabus as "good for students" and "helpful for advisors" and "support services."

Staffing efforts began with a deliberate and difficult argument that we would need individuals with training and expertise in rhetoric and composition if we wanted to significantly improve our program. We (the chair and I) drew on the distrust in FYC delivery in order to make arguments for lines, credentialing, and professional development/evaluative review for the faculty teaching our courses. In addition, we drew on campus-wide attitudes toward TA labor to make an argument for required pre-semester training and sustained requirements for professional development for all TAs.

This argument was successful and resulted in a much more robust approach to training and supervising our graduate teaching assistants. I also worked to offer a variety of professional development opportunities for all writing program instructors, focusing almost exclusively on research from writing studies and even bringing a very well-respected writing studies scholar to campus. Essentially, I strategically drew on the dominant discursive threads of mistrust and used them to my advantage by subverting them for my own purposes. These afforded my program some gains, yes, but I am less sure whether it resulted in anyone outside my department acknowledging my "claiming" of writing studies as my discipline.

This shift also resulted in a restructuring of our master's program, adding in an intentional track for studying rhetoric and composition, thus increasing the number of graduate course offerings in rhetoric and composition (beyond just the teaching practicum). I also increased observations and mentoring, along with several policies aimed at consistency of delivery across FYC courses. In some ways, all these efforts at increased supervision and standardization appeased external stakeholders but were met with some internal resistance (no surprise to my readers here, I'm certain). My management of such resistance is and has always been to listen authentically, make changes based on recommendations and inclusive voting, and create intentional opportunities for contributing to curriculum and policy redesign.

While I do see several of these efforts to innovate curriculum and improve staffing as successful enterprises, my ability to tackle the WAC-lite phenomenon at my current institution presented a much thornier challenge. I recognized that I would need to position myself advantageously beyond the department if I hoped to advance any WAC/WID initiatives grounded in research from writing studies. To that end, I volunteered to serve on the general education deliberative body on my campus, and eventually became chair of the core curriculum committee governing writing courses and the elected chair of the entire council. In this role, I advocated continually and persistently for the impact of vertical writing instruction, and within eighteen months I successfully secured a majority vote for the addition of an upper-division writing requirement. With this addition to the general education curriculum, I was able to draw on the state articulation of "content criteria" and "competencies" to define the upper division requirement as a discipline-specific writing requirement. I knew that such a move would allow me to draw on a more powerful voice than my own (the State) in order to advance a WAC/WID approach to upper division writing requirements on my campus. Moreover, I recognized that because the un-anchored WAC-lite was happening in the second course (of the now three-course sequence), the newly approved addition would force a conversation about whether or not those courses can or should be moved to the upper division and, if so, who (or what body) would determine whether or not all these new upper-division courses were consistently adjudicated and delivered. During this same period of time, I created a key alliance with the campus faculty development center in order to develop a grant initiative aimed at incentivizing and supporting faculty interested in developing WAC/WID courses.

When reflecting on the rhetorical and strategic maneuvering outlined above, I find myself torn on whether it represents a story of triumph or a cautionary tale. In the edited collection *Kitchen Cooks, Plate Twirlers, and Troubadours: Writing Program Administrators Tell Their Stories*, Diana George (1999) writes that "it is through such stories, tied to the scholarship, research, and teaching that continue to shape our profession, that we mentor each other and see beyond self" (180). While I do see my efforts to orchestrate the advancement of an emergent WAC/WID program at my institution as a success, I continue to struggle with convincing faculty across campus of the value of WAC/WID professional development training being couched in my discipline of writing studies. Successfully integrating writing studies in our FYC courses is an easier task (those courses were ours to begin with). Successfully positioning WAC/WID course development in writing studies is more challenging. The perception is that upper-division courses are grounded in the discipline in which they are offered. And, of course, that's correct. And yet, considerable scholarship, theories, and best practices for WAC/WID pedagogy and course design emerge from the discipline of writing studies.

Moreover, it's important to point out that I work at an institution where no WAC or WID program currently exists (not in practice or infrastructure). In my fervor to anchor writing instruction in writing studies expertise, have I just signed myself up as both WPA and WAC coordinator? This represents yet another risk of claiming writing studies. I now face the challenge of making an argument for an additional faculty administrative line.

Furthermore, in my efforts to get the upper-division writing required voted through, have I relied too heavily on a notion of discipline-based writing that advances the neoliberal reconstructions of the university? In the introduction to the edited collection *Writing against the Curriculum: Anti-Disciplinarity in the Writing and Cultural Studies Classroom*, editors Randi Gray Kristensen and Ryan Claycomb (2010) look closely at a description of WID in a business school to "see WID's potential complicity with the corporatization of the university and its use of writing as an instrumentalist component of the production of labor capital for the workforce" (11). They go on to claim that "the process of disciplinarity compartmentalization is linked directly to transactional writing designed specifically for use in the marketplace" (Kristensen and Claycomb 2010, 12). Intentionally integrating the home discipline of a WAC/WID course with research from writing studies could potentially offset some of this emphasis on a transactional approach to writing instruction. Reflecting on my success as a WPA requires me to think as intentionally as possible about risk and reward, gains and losses. Twirling plates indeed.

Assessment, Retention, and Social Justice

As is the case at many institutions, my campus is deeply concerned with improving the retention and persistence rates for our students. My institution has a strong representation of first-generation students and is also currently looking to secure Hispanic-Serving Institution (HSI) status. Our retention rates are low, and several initiatives are currently underway to improve our student success. As a relative newcomer to this institution, I knew that my experience in WPA work could position me well to contribute to such efforts. I also knew within a short time after my arrival that the mistrust of English offerings in general education writing courses was profound.

In a quantitative study examining the importance of FYC course grades in predicting retention and persistence, Garrett, Bridgewater, and Feinstein (2017) found that "performance in writing courses strongly predicts both graduation and success in the major" (107) and that "among all first-year GE [general education] courses, failing a writing course is the best indicator of a student not graduating" (107). Ultimately, they argue that "for institutions focused on improvement, dealing with high-dropout courses with highly predictive natures may be an ideal way to identify and resolves deficiencies" (108). While my localized data did not show our courses were "high-dropout courses," I was able to draw on institutional research to demonstrate the highly predictive nature of our first-year writing courses. This data provided me with a talking point with student success administrators. I sought out the opportunity to participate in strategic planning committees focused on student success, and as a result of my involvement on such committees, we implemented an Early Performance Feedback pilot program in a sample of first-year writing courses. This pilot draws on research about intervention for students who present with early indicators of retention-risk (e.g., not having a textbook or missing more than two days of class in the first two weeks). Thanks to this pilot in the writing program, my institution is now considering how to scale this project to the larger campus. My involvement positioned the writing program as a leader and innovative hub on campus, especially in regard to issues of retention and persistence. While this status has allowed me to make more effective arguments across campus for resources and buy-in, I am left to wonder which served my purposes more: the nature of my institutional position as WPA or my sharing disciplinary research. Perhaps both.

In critically reflecting on this success, I have been drawn to Rita Malencyzk's (2017) article, "Retention ≠ Panopticon: What WPAs Should Bring to the Table in Discussion of Student Success." She tells two stories to open her piece, one about her son's college experience and one about a retention initiative on her own campus, where she serves as WPA. She argues that "implemented uncritically, retention efforts can turn a university into a panopticon, a Foucauldian instrument of power and control that reflects a fear of the disorder that often accompanies human agency and human development, particularly the development that occurs in the adolescence of traditional-aged college students"

(25). In claiming status for writing studies and my writing program, am I complicit in panoptic retention efforts? Of course, not all retention efforts are created equal, and some are perhaps more guilty of Malencyzk's critique than others, but claiming writing studies requires that as we advance our programs and as we aim to gather support and resources for our instructors and our courses, we must also acknowledge the challenges our discipline has offered from within.

I have also experienced this same tension in my assessment work. With several years of assessment experience under my belt, I sought out a grant-funded opportunity to assess the first course in the writing program at my current institution. However, having been heavily influenced by social justice work in writing studies (Inoue 2015; Poe, Inoue, and Elliot 2018), I found such work at odds with (or at the very least constrained by) the state's articulation of competencies, content criteria, and the evaluative rubric they recommended for assessment purposes. And yet, I knew that if I could demonstrate an assessment of our students' achievement relative to the state learning outcomes, not only would that likely garner praise among upper administrators, but it may also serve to help me further anchor all writing instruction courses across campus. Thus, a tension remains for me regarding how to advance what writing studies has so persuasively argued regarding racist assessment practices (within the classroom space) while also administrating under state constraints.

Concluding Thoughts

In reflecting on my work as an administrator and on my efforts to claim writing studies as my discipline, I recognize that claiming it strategically is imperative and that even my failures can be reflected upon as "a useful lens for analyzing the starts, stops, and missteps" that have occurred along the way, as Brooks, Dadas, Field, and Restaino explain in chapter 4 of this volume. Identifying the when and where to make these strategic declarations is critical when balancing one's own passion with the existing ideologies and constraints that persist in higher education. Developing kairotic and intentional partnerships has afforded me some success in advancing the mission of the various programs I have had the privilege to administer. I am also mindful, however, that alongside these strategic efforts, we must also prioritize what we can manage and at what pace. Embodying Micciche's (2011) slow agency requires mindful, critically reflective decision-making about what comes first, second, and fiftieth. What does it mean to claim our disciplinarity? What risks come with such declarations? And, of course, these declarations are not monolithic. What I count as cornerstones of our discipline may differ significantly from another writing studies specialist. Scholars in our field are hard at work addressing the nuances of claiming disciplinarity. Reconciling our internal articulations of best practice with our daily goals is a project most all WAC/WID directors or WPAs wrangle with daily, in our meetings, in our scholarship, and in our programs. Gere et al. (2015), for example, offer the concept of *new disciplinarity*, troubling static disciplinary conceptions with concepts such as borderlands, temporality, and elasticity. In similar fashion, writing studies scholars such as Trader et al. (2016) tackle the challenge of maintaining disciplinary standards under tremendous pressure from external political initiatives. We have to learn from and listen to our field's efforts to claim disciplinarity in the face of a myriad of barriers to such recognition. We have to share our administrative strategies, our rhetorical tactics, and our stories of resilience.

Note

1. My institution is a public, regional doctoral-granting institution in the western United States, serving approximately twelve thousand undergraduate and graduate students. Nearly half of the students are first-generation, and the university is currently seeking to gain Hispanic-Serving Institution (HSI) status. The writing program at this instruction is guided by state requirements and content/outcomes articulations. The general education program includes a requirement for six credit hours in written communication courses. There are three pathways available to students for completing this requirement: (1) completion of an introductory composition course with supplemental academic instruction and completion of an intermediate composition course, (2) completion of an introductory composition course and completion of an intermediate composition course, or (3) completion of an intermediate composition course and an advanced composition course.

References

Adler-Kassner, Linda, and Elizabeth Wardle. 2015. *Naming What We Know: Threshold Concepts in Writing Studies*. Logan: Utah State University Press.

Adler-Kassner, Linda, and Elizabeth Wardle, eds. 2019. *(Re)Considering What We Know: Learning Thresholds in Writing, Composition, Rhetoric, and Literacy*. Logan: Utah State University Press.

Balzhiser, Deborah, and Susan H. McLeod. 2010. "The Undergraduate Writing Major: What Is It? What Should It Be?" *College Composition and Communication* 61, no. 3 (February): 415–433.

Banks, Adam. 2015. "Funk, Flight, Freedom." CCCC Chair's Address, Tampa, Florida, March 24, 2015.

Bazerman, Charles. 2011. "The Disciplined Interdisciplinarity of Writing Studies." *Research in the Teaching of English* 46, no. 1 (August): 8–21.

Garrett, Nathan, Matthew Bridgewater, and Bruce Feinstein. 2017. "How Student Performance in First-Year Composition Predicts Retention and Overall Student Success." In *Retention, Persistence, and Writing Programs* edited by Todd Ruecker, Dawn Shepherd, Heidi Estrem, and Beth Brunk-Chavez, 99–113. Logan: Utah State University Press.

George, Diana, ed. 1999. *Kitchen Cooks, Plate Twirlers, and Troubadours: Writing Program Administrators Tell Their Stories*. Portsmouth, NH: Heinemann-Boynton/Cook.

Gere, Anne Ruggles, Sarah C. Swofford, Naomi Silber, and Melody Pugh. 2015. "Interrogating Disciplines/Disciplinarity in WAC/WID: An Institutional Study." *College Composition and Communication* 67, no. 2 (December): 243–266.

Heilker, Paul, and Peter Vandenberg. 2015. *Keywords in Writing Studies*. Logan: Utah State University Press.

Inoue, Asao. 2015. *Antiracist Writing Assessment Ecologies: Teaching and Assessing Writing for a Socially Just Future*. Fort Collins, CO: WAC Clearinghouse.

Isaacs, Emily. 2018. *Writing at the State U: Instruction and Administration at 106 Comprehensive Universities*. Logan: Utah State University Press.

Kristensen, Randi Gray, and Ryan M. Claycomb, eds. 2010. *Anti-Disciplinarity in the Writing and Cultural Studies Classroom*. Blue Ridge Summit, PA: Lexington Books.

Macmillan Learning. 2014. "Meet the Authors of Writing about Writing." YouTube. November 1, 2014. https://youtu.be/UDyPE2IMD68.

Malencyzk, Rita. 2017. "Retention ≠ Panopticon: What WPAs Should Bring to the Table in Discussion of Student Success." In *Retention, Persistence, and Writing Programs*, edited by Todd Ruecker, Dawn Shepherd, Heidi Estrem, and Beth Brunk-Chavez, 21–37. Logan: Utah State University Press.

Malencyzk, Rita, Susan Miller-Cochran, Elizabeth Wardle, and Kathleen Blake Yancey, eds. 2018. *Composition, Rhetoric, and Disciplinarity*. Logan: Utah State University Press.

Micciche, Laura. 2007. *Doing Emotion: Rhetoric, Writing, Teaching*. Portsmouth, NH: Boynton/Cook.

Micciche, Laura. 2011. "For Slow Agency." *Writing Program Administration* 35 (1): 73–90.

Phelps, Louise Wetherbee, and John M. Ackerman. 2010. "Making the Case for Disciplinarity in Rhetoric, Composition, and Writing Studies: The Visibility Project." *College Composition and Communication* 62, no. 1 (September): 180–215.

Poe, Mya, Asao B. Inoue, and Norbert Elliot, eds. 2018. *Writing Assessment, Social Justice, and the Advancement of Opportunity*. Fort Collins, CO: WAC Clearinghouse.

Robertson, Liane, and Kara Taczak. 2018. "Disciplinarity and First-Year Composition: Shifting to a New Paradigm." In *Composition, Rhetoric, and Disciplinarity*, edited by Rita Malenczyk, Susan Miller-Cochran, Elizabeth Wardle, and Kathleen Blake Yancy, 185–205. Logan: Utah State University Press.

Rose, Shirley K. 2004. "Representing the Intellectual Work of Writing Program Administration: Professional Narratives of George Wykoff at Purdue, 1933–1967." In *Historical Studies of Writing Program Administration: Individuals, Communities, and the Formation of a Discipline*, edited by Barbara L'Eplattenier and Lisa Mastrangelo, 221–239. Anderson, SC: Parlor Press.

Rose, Shirley, and Irwin Weiser, eds. 1999. *The Writing Program Administrator as Researcher*. Portsmouth, NH: Boynton/Cook.

Scott, J. Blake, and Lisa Meloncon. 2017. "Writing and Rhetoric Majors, Disciplinarity, and Techne." *Composition Forum* 35 (Spring). https://compositionforum.com/issue/35/majors.php.

Trader, Kristen Seas, Jennifer Heinert, Cassandra Phillips, and Holly Hassel. 2016. "'Flexible' Learning, Disciplinarity, and First-Year Writing: Critically Engaging Competency-Based Education." *WPA: Writing Program Administration* 40 (1): 10–32.

Vidali, Amy. 2015. "Disabling Writing Program Administration." *WPA: Writing Program Administration* 38, no. 2 (Spring): 32–55.

Wardle, Elizabeth, and Doug Downs. 2016. *Writing about Writing*, 3rd ed. New York: Bedford.

Yancey, Kathleen Blake. 2018. "Mapping the Turn to Disciplinarity: A Historical Analysis of Composition's Trajectory and Its Current Moment." In *Composition, Rhetoric, and Disciplinarity*, edited by Rita Malenczyk, Susan Miller-Cochran, Elizabeth Wardle, and Kathleen Blake Yancy, 15–35. Logan: Utah State University Press.

Yancey, Kathleen Blake, Liane Robertson, and Kara Taczak. 2014. *Writing across Contexts: Transfer, Composition, and Sites of Writing*. Logan: Utah State University Press.

4
Embracing Failure

A Newly Independent Department's Attempts at Writing Its Own Script

RON BROOKS, CAROLINE DADAS,
LAURA FIELD, AND JESSICA RESTAINO

Introduction

In the fall of 2016, as the result of departmental conflicts that were deemed irreconcilable by an administration that was inclined to support a separate department, a new Department of Writing Studies broke off from the English Department at Montclair State University. For one semester, this department was housed directly within the Dean's Office but was granted "independence" in the spring of 2017. At the time, the department consisted of tenured faculty member Caroline Dadas, who served as interim chair; tenured faculty member Jessica Restaino, who served as director of First Year Writing; and twenty-eight full-time instructional specialist faculty, who primarily taught first-year writing courses. Of these twenty-eight instructional specialist faculty, only two had degrees in composition and rhetoric. The department also employed a considerable number of part-time adjunct faculty. All of the faculty (including the chair) are unionized, though the adjunct faculty are part of a different union than the full-time faculty. This same semester (spring 2017), the department hired two full-time tenure-track faculty members, one of whom was Ron, our founding chair.

https://doi.org/10.7330/9781646426331.c004

We begin with these details to set the stage for various stories of how "claiming our own" as a department has been met with many complicating tensions. First, there are residual tensions from the break with the English Department. Second, there are numerous complications surrounding the various service loads held by the full-time tenure-track faculty in our department. Finally, and most importantly for our purposes, the unequal status and contracts of the full-time instructional specialist faculty puts limitations on many of the collaborative ideals that the department would like to perform. We address all of these topics later in the chapter, but first we offer an overview of the methods and methodology that we are employing.

Collaborative Autoethnography and Queerness

In March 2019, Ron Brooks, Caroline Dadas, Laura Field, and Jessica Restaino presented "A Newly Independent Department Writes Its Own Script" at the Conference of College Composition and Communication, reflecting on how our new departmental affiliation has given us the opportunity to institutionally perform core beliefs of writing studies such as valuing collaboration, working toward more fair labor models, and engaging in interdisciplinary curricular development. We reflected on the departmental separation and also unpacked what has been made possible since writing studies has become its own departmental entity at Montclair State. Specifically focusing on how creating a new department has offered *kairotic* opportunities for practicing social-justice-minded pedagogies, administrative structures, and labor models, we described the rhetorical tactics we have employed in order to define our discipline.

Expanding on this initial project, Caroline Dadas and Ron Brooks decided to rework this initial presentation into a full-length collaborative autoethnography. With Jess Restaino's commitment to new initiatives in her directorship of the Gender, Sexuality, and Women's Studies program and Laura Field's ongoing institutional work to secure better contracts for all faculty (including instructional specialists), their participation took the form of reading drafts and providing feedback as the project progressed. By expanding this initial presentation into a longer essay, the authors hoped to bring this work into conversation with the work that Everett and Hanganu-Bresch (2017) brought to light in *A Minefield of Dreams*, a book that was published right at the time that their department was formed.

Continuing this conversation seemed vitally important to all four of us because of our hope to focus more specifically on the implications of an independent writing program that is explicitly committed to aspects of social justice. Not only is our department committed to social-justice-oriented pedagogies but we as authors of this chapter want to express a methodological commitment to social justice as well. For this reason, we are employing a queer methodology, which we believe offers an important contribution to the literature around independent writing departments. We draw on queer methodologies because they offer us much in the way of thinking through how norms are established, circulate, and can be disrupted. For example, reflecting on the "well-trodden path," Sara Ahmed (2006) writes in *Queer Phenomenology*, "Lines are both created by being followed and are followed by being created. The lines that direct us . . . are in this way performative: they depend on the repetition of norms and conventions, of routes and paths taken, but they are also created as an effect of this repetition" (16). In the very makeup of our department, we found ourselves positioned to disrupt this repetition of norms, conventions, and routes taken—though it meant encountering resistance to decades of established practices. Throughout this chapter, we explore how queer methodologies and approaches—particularly the concepts of failure and queerness as *techné*—helped us negotiate our roles as administrators and address some of the many challenges that continue to confront our new department. We will provide a brief overview of these concepts before moving on to our narrative background and analysis.

The reframing of the concept of failure by queer theory scholars offers us a useful lens for analyzing the starts, stops, and missteps that have accompanied our transition to an independent department. Jack Halberstam (2011) envisions failure, despite its connotations, as a productive practice, one that disrupts or queers expectations, assumptions, or doing things a particular way simply because we've always done them that way. In essence, failure can represent one way of productively shaking up norms. Halberstam argues that "queer studies offer us one method for imagining, not some fantasy of an elsewhere, but existing alternatives to hegemonic systems" (89). The hegemonic system in question here is the current labor model of the modern-day American university, one in which a multitiered system is firmly in place. Our departmental structure in which instructional specialist faculty represent a majority offers an opportunity to queer accepted notions of what the relationship between tenure-track and full-time faculty can look like.

Along with failure, we also find the concept of queerness as *techné* helpful in thinking through the particulars of our new department. Caroline has written previously about how queerness embraces messiness, rejects categorization, and refuses binaries. Considering queerness as *techné* (theoretical knowledge combined with practical skill) helps us view the change that we experience as a new department as an opportunity for growth, neither good nor bad. For Caroline, "viewing queerness as *techné* helps us to reorient toward the process of adaptation, the flexibility of method, the need to constantly change our approaches" (Dadas 2016, 70). We find this framework useful from our current administrative vantage points. As the department matures, we need to remain flexible and adaptive, not adhering to previous scripts about what an effective department or first-year-writing program should look like. In fact, some of the recent changes that we've experienced open opportunities for faculty to reorient our focus. For example, we have a new major in public and professional writing (PPW) that needs to be promoted and developed; faculty might now turn greater attention toward recruitment for PPW courses, a need that did not previously exist. "Imagining . . . existing alternatives to hegemonic systems" can mean refusing to give in to binaristic thinking; that changes are either good or bad.

For the process of composing this chapter, we followed a method of collaborative autoethnography because it allows multivocality to be prominent in the piece. We find this trait important given that we each occupy different vantage points in the department and come to the current moment through different paths (for example, Jess, Laura, and Caroline experienced the separation process from English, while Ron joined Montclair State after this process had been completed). Collaborative autoethnography gives us a qualitative method that allows us to reflect on our own experiences in a structured (but not prescribed) way.

We began writing this chapter by using our collective Conference on College Composition and Communication presentation as a foundation, and then we began to build on and reshape this text. At three different junctures in the revising process, Caroline and Ron met to write individual reflective memos based on the text produced so far and related readings. We would read each other's memos, and then use the Google Docs comment feature to respond to the other person. When we would meet for our next writing session, we would repeat the process. We then sent drafts of the chapter to Jess and Laura, who provided their own feedback and input, which we incorporated. This recursive approach guided our collaboration and also informed our thinking about our own department as we moved forward. Through this chapter, we are writing ourselves into departmental identity. Given the whirlwind of our separation from English and the ensuing work needed to get the new department off the ground, this written effort represents our first sustained attempt at reflecting on where we now find ourselves as a department. Our collaborative and non-hierarchical approach to writing this piece is what we seek to establish in our department; below we assess where we are in the process of establishing this kind of work environment.

Four Different Ways of Looking at the Founding of a Department

The following are four pieces, adapted from our initial CCCC presentation, which help provide further context for the stage set in our introduction. It is our hope that these particular stories will help others become more fully attuned to the *kairotic* moments that surround department formation. These stories are important because they provide fuller context for the decisions that Ron and Caroline wrote their way into in later portions of the essay. They also underscore how messy the process of writing one's way into a departmental vision can be.

JESS

When we were part of the English Department, we had a long history of struggling to get curricula passed. Typically, we would eventually be successful, but we often had to fight for the legitimacy of the field and would face questions that undermined the notion of our field as a disciplinary entity. By the spring of 2015, we had a handful of undergraduate courses in writing studies (and a nascent minor in public and professional writing). We also had a healthy MA concentration.[1] We were aware that enrollments in the literature-focused English major were on the decline, in line with national trends. After privately discussing our frustrations with the current state of the English Department, we realized that we were in the midst of a *kairotic* opportunity. Our college had just hired a new dean, who on several occasions had mentioned his support for writing studies programming. Additionally, the English Department would undergo a five-year external review that year (the 2015/2016 academic year). We realized that any action that we took on behalf of the discipline would have to be addressed by the external committee, and we felt confident that they would support our initiative to grow writing studies within the department.

With that context in mind, we (Jess and Caroline; Laura was on a contingent contract and thus more vulnerable to repercussions) proposed the following: an undergraduate writing studies major as a way to help support overall departmental enrollments, and two writing studies tenure-track lines to support our already-existing needs (Caroline taught the majority of our PPW courses at that time because she had a 3/3 teaching assignment). Both proposals were met with defensiveness and, in some cases, anger. We even had difficulty getting on the agenda for the curriculum committee, and when we finally did, we faced accusations that we were going to "steal" students from literary study. The committee asked us to gather data to prove that the writing studies major would not draw students away from the literature-focused major. Discussions were eventually tabled when the English Department chair put forward her own proposal to revise the English major. When it came to the tenure-track faculty lines, Caroline and I were the only two faculty who voted to move them forward (instructional specialist faculty, despite teaching the majority of the classes in the English Department, did not have voting rights).[2]

While we were met with a negative reception from our English Department colleagues, both the external review committee and the dean were supportive of our curricular plans. In the spring of 2015, the English Department wrote a response to the external committee that indicated they would not move forward on the committee's recommendations to further develop writing studies within the department. A few months later, the dean announced the creation of the Department of Writing Studies. Although many factors were beyond our control, and the result was never certain, Jess and Caroline accurately read the departmental landscape in 2015 as one that was *kairotic* for change.

LAURA

My position at the university is an instructional specialist faculty position—a different position than Jessica, Caroline, or Ron. I am one of only twenty-eight instructional specialist faculty in our department, so I am sharing with you my personal experiences, which may not reflect the exact experiences or perceptions of all my colleagues.

Until 2013, or about two years prior to our split from English, our full-time contracts only allowed for three continuous years of employment at one time. This stipulation meant that once our contracts ended, we needed to take one year teaching at 3/4 time, hoping to be hired back the next year on another 4/4 full-time contract. Despite this precarity, we operated in our own little bubble within the English Department, which felt like a safe space. We had committees, designed curriculum, and contributed to policy discussions—all under the radar of the larger department.

In 2013, the English Department began hiring into the institution's newly created instructional specialist positions. Although far from perfect, I believe this new position led to a shift in the way full-time faculty, including myself, viewed our roles at the university: we saw these positions offering at least the possibility of longer term employment. We began to think about how we fit in the department and the university. In these precarious positions, there is desire to feel seen and validated—that your contributions to university have worth. There is also a desire to be involved in order to help gain some sort of stability or promise of employment, but there is also the risk of exploitation and rejection. We were trying to strike a balance between becoming more fully integrated into the life of the university and not doing work that left us feeling undervalued.

By 2015, we encountered growing resistance from the English Department to our new roles. As our numbers grew to twenty-eight instructional specialist faculty, the department began taking steps to assure that the tenure-track faculty would not be "outnumbered" when it came to governance of the department. In other words, the number of full-time writing studies faculty in the department now rivaled the number of faculty who specialized in literature, film, and creative writing combined. As more instructional specialist faculty began attending department meetings, the department opened a discussion about whether we could vote. After several contentious meetings during which we were portrayed as imposters, the department decided that we would not have voting rights. Another major sticking point arose when the department began revising its bylaws and needed to decide how they would refer to us in this document. The union contract forbade that we be called faculty, as a way of distinguishing between instructional specialist faculty and tenure-track colleagues. The instructional specialist faculty met privately and decided on wording that would refer to us as "members" of the English Department. This suggestion was rejected, and we were essentially erased from the departmental bylaws, despite the fact that we were collectively teaching approximately 224 sections per academic year in that department.

As a result, we began discussions about how we should let the English Department leadership know that we felt disrespected and marginalized. An opportunity soon presented itself: The department chair scheduled a group photo of the tenure-track faculty, and she invited the instructional specialist faculty by email to be included in the picture.[3] We viewed this gesture as a surface-level attempt at appearing inclusive, while excluding us from all meaningful aspects of departmental governance. The instructional specialist faculty agreed unanimously to boycott the photo, and we drafted a statement that detailed our reasons why. We planned to email the statement to the entire department. The group decided that two of us, including myself, would schedule a meeting with the chair in advance as a courtesy. After a brief meeting with the chair, we arrived back to our offices to see that she had already emailed a message to the department saying that the photo session had been canceled, without explanation. We viewed this move as an attempt at undermining our *kairotic* opportunity. At the time, it felt like we had failed at what we had hoped to be a meaningful gesture of protest. In retrospect, this event seems to fit the overall pattern at that time where any attempt to integrate the instructional specialist faculty into the norms of the department revealed that we were not a good fit for what the department desired. Repeated failure in this sense became a clarifying process for all parties, even if we knew the road ahead (i.e., leaving the department) would be bumpy. We sent the email anyway, forcing the departmental leadership to acknowledge and address this deteriorating situation. At the same time as these events were transpiring, Caroline and Jess were advocating for curricular change and tenure-track lines. Soon, the possibility of separation would be broached by the upper administration, with all of these occurrences standing as evidence for why the status quo was no longer tenable.

CAROLINE

As one of the two tenured professors who moved from the English Department to the Department of Writing Studies in January 2017, my interim chair position proved a *kairotic* opportunity for observing how non-tenure-track faculty are marginalized within the broader campus community. The treatment of the full-time faculty who taught First-Year Writing (contractually termed "Instructional Specialists") as less-than-full members of the English Department represented a central precipitating cause for separation. Once the separation became final, the new department then had to forge new institutional ground in this area. Because the department was established in the middle of an academic year, I was asked to serve as the interim chair from January to August 2017, while we hired our founding chair (Ron). My new position opened up opportunities to contest the marginalization of the instructional specialist faculty who moved from the English Department to the Writing Studies Department—or, at the very least, to call attention to this marginalization, and publicly question some unspoken norms publicly. Our ratio of full-time faculty (28) to tenure-track faculty (4) within our department marked us institutionally, from the beginning, as aberrations.

One of our first tasks as a new department was to establish a committee structure. Not only did we have to develop departmental committees but we also had to decide on representatives to college-level committees. One representative whom we needed to appoint was to the college-wide curriculum committee; this person would vote on the curricular proposals of other departments and would be responsible for bringing our own curricular proposals to the group. An instructional specialist faculty member indicated interest, and she was elected to the position (this person holds a PhD in writing studies). After her first meeting on the committee, I received an email from the chair of the committee saying that a fellow committee member had raised objections to the fact that our representative was not on the tenure track. The bylaws of this committee stated that only tenure-track or tenured faculty could serve on the committee, though these bylaws were written before the current instructional specialist positions were created.

In doing some research about who currently sat on the committee, I discovered that an instructional specialist faculty member had been participating on the committee for years. When I brought my concerns back to the committee chair, she offered me the opportunity to deliver a statement at the next meeting. Prior to the meeting, I asked for both the elected representative's input (we agreed that she would not attend the meeting) and Laura's input about rhetorical approaches for my statement; although I was the faculty member who would be present, I wanted to make sure that multiple instructional specialist faculty perspectives were taken into account and that these faculty would have some measure of representation at the table.

In my remarks, I offered the committee a rationale for why they should allow our representative to serve. Afterward, a discussion ensued in which several faculty claimed that instructional specialist faculty have no expertise in curriculum, they are not sufficiently trained/credentialed, and they are only interested in "taking" tenure-track positions. In addition to these comments, several faculty on the committee sincerely asked "who these folks are," illustrating a pervasive lack of understanding of how labor is distributed at our university. The committee chair followed up with me after the meeting to say that the committee would have to discuss whether they wanted to alter their bylaws to accommodate instructional specialist faculty, and until then, they could not allow our representative to serve. I stepped in as the representative for the remainder of the academic year.

In reflecting on these experiences, I see them as illustrative of a queer conception of failure, as articulated by Jack Halberstam (2011) and others. Ultimately, Caroline's arguments were not successful in the traditional sense of the word: the committee was not willing to allow our elected representative to serve. But because of our department's actions, a dialogue about instructional specialist faculty and their role within the larger university structure was initiated in this space. In asking the committee to seat an instructional specialist faculty member on the committee, we were further trying to complicate the division between tenured and not tenured faculty. Because of our request, the *potentiality* of having instructional specialist faculty participation on college-wide committees has now been broached. We are drawing the term *potentiality* from José Esteban Muñoz (2009) who, in *Cruising Utopia*, argues that "unlike a possibility, a thing that simply might happen, a potentiality is a certain mode of nonbeing that is eminent, a thing that is present but not actually existing in the present tense" (9). Potentiality serves as a form of critique by imagining how things might be different. Even though our department did not achieve success in the short term, we like to think that this kind of work, which surfaces norms and instigates difficult conversations, is required in order for more equitable practices to eventually come about. We directly challenged what is and what is not permissible, laying the groundwork for potential change in the future. In this scenario we find resonances with Karen Tellez-Trujillo's description of feminist resilience in her chapter of this volume: "Continuous forward movement . . . progression through consistent, small gestures." What matters is the forward movement: not significant leaps, but also, as Tellez-Trujillo notes, not a return to a prior state either. While the labor model of the university encourages the tenure track and non-tenure track relationship to be oppositional, we implicitly asked what other models are possible, creating forward motion around that question.

RON

In the latter part of 2016, when I was first interviewed for the position of chair of Writing Studies at Montclair University, the first question I was asked was, "Why do you want to be our founding chair?"

Desire is a funny thing. For the twelve years I had been in an English department in a research university in the Midwest, I had fantasized about creating a freestanding department. I imagined in this department not being held subservient to literature, screen studies, creative writing, linguistics, and so on. I imagined not having to argue with people who did not see composition, rhetoric, or technical writing as real disciplines. I imagined not having to continually fight the battle within my department for new hires.

But perhaps most importantly I wanted to lead a department using a consensus model. My previous department had in place a head system, where the department head is contractually defined as administration rather than faculty, and being the associate head of that department, it was clear that OSU's structure did not encourage a consensus model of leadership, even though I had seen some examples of consensus leadership while there. But for me, even down to the metaphor, being a chair (instead of a head) meant that I would be sitting at a table with my colleagues. For me, joining (and leading) an independent writing program would allow me to claim what I discern is a key part of the discipline of writing studies: the dissolution of traditional academic hierarchies.

But I had concerns about making the move to a freestanding writing program. As outlined by Justin Everett and Christina Hanganu-Bresch (2017), writing studies programs continue to wrestle with whether they should market themselves within the rhetorical tradition of the liberal arts or emphasize the more practical tradition of rhetoric, and even in my first interviews with Montclair State it seemed that tension was a source of dissonance for the department.

At the time, though, my concerns about moving were less theoretical and more practical. At that time, the English department where I worked could often be a contentious place, but it was a large department that was not used to being pushed around by upper administration. That is to say, as a comp/rhet person, I discovered that if I could get something through my department it would pass through the dean and provost with very little pushback. As my rhet/comp colleagues who worked and continue to work there would likely tell you, though, there were many rhet/comp initiatives that did not make it past the departmental stage.

Montclair State at that time had four tenure-track faculty, twenty-eight full-time instructional specialist faculty, and a varying body of adjunct faculty. Departmentally, it had a vibrant, award-winning First Year Writing program, an established and growing minor, and only the promise of a major. I was concerned that it would be in a position to be pushed around in ways that a larger program could not. But going to Montclair's campus and getting to know both the tenure stream and instructional specialist faculty made it clear to me that I wanted to play a part in this new department's self-definition. And perhaps more important than the way that the department would define itself theoretically was the clear fact that this department wanted to define itself as a model of collaboration.

It was not clear to me at the time the degree to which the ratio of full-time non-tenure-track faculty (28) to tenure-track faculty (4)—not to mention the high number of adjunct faculty (50+) who were teaching in our department—would complicate that particular ideal, and so my reflection on my experience comes from a different place than Caroline's. With that said, I have been fundamentally changed by Caroline's articulation of a queer conception of failure, as it has opened up new ways for me to think about the potentialities that emerge from places of eminent non-being, and for me a great deal of that potential would reside in the way that our department publicly moved forward with its bylaws and its major. The experience has also opened up for me avenues of thinking that would not have been possible for me when I was a graduate student first studying composition and rhetoric in the early 2000s. Even though there were a few independent writing programs, and Peggy O'Neil, Angela Crow, and Larry Burton (2002) had published *Field of Dreams: Independent Writing Programs and the Future of Writing Studies*, the full implications of these relatively new programs had not yet been fully hammered out on the anvil of experience.

Early Attempts (and Failures) at Claiming Our Own

Having set the groundwork for our thinking about the early stages of our department, we will now articulate the results of our recent reflections. After reading back through our presentation and writing together, we (Caroline and Ron) eventually arrived at the following research questions:

- How has our perspective as a department changed in the years since its founding?
- Has our experience altered our ideals? How?
- How do our standpoints, our identities, and our current administrative roles shape who we are and how we function in the department?
- What are the implications of independence in the above?

What follows is the writing and insights that emerged through this process.

RON: YOU STILL HAVE TO SERVE SOMEBODY:
BYLAWS AND THE LIMITS OF COLLABORATION

Early on, I saw one of the big tests of our ideal of collaboration within the framing of the structure of our department in our bylaws. How would faculty be named, who would have the power to vote? It was clear that our department wanted to include all instructional specialist faculty, but what about the power of adjunct faculty?

On paper, I would count our bylaws as a success. We came to an agreement that gave full-time instructional specialist faculty the power to vote in our department and perhaps subversively the power to be called "faculty" in our bylaws. We also, and to some degree more controversially, gave adjunct faculty the power to vote at our department meetings, and put into place a structure that would inform adjunct faculty when there would be issues that would most directly affect them.

For me, this was far different than any structure I had previously seen. We do not have graduate students teaching first-year writing, so the conflicts that at times keep faculty and their graduate students separate from each other were not a factor. It was entirely possible to have everyone teaching in our department present for our decisions.

Where "failure" or "the mess" enters the equation for me is when we move away from what happens on paper to actual practice. I was naïve, I now believe, in some of my thinking about collaborative leadership. As Caroline wrote to me in one of our writing sessions,

> I wonder how having a large number of Instructional Specialist faculty in the department has influenced / complicated [your collaborative] approach. In most departments at MSU and elsewhere, the chair is a colleague of the full-time faculty in the department. The chair has some authority, to be sure, in setting an agenda, overseeing scheduling, etc.—but the chair is not the boss. In our department, the chair is the boss of everyone except for 3 other people. How does that impact the consensus model?

It most definitely complicates the consensus model, particularly now that our department has successfully gotten its major passed through curriculum committees and an incredible amount of resistance and redefinition on the part of the upper administration.

The major, it is now clear to the department, does not have literature classes in it. This possibility was always clear to instructional specialist faculty with backgrounds in rhetoric and composition and writing studies, but we now realize this was perhaps not clear to all of the faculty in our department. When coupled with Caroline's recent efforts to refocus our first-year-writing curriculum, the resistance toward our being an actual writing studies department has intensified and has made it clear to me how difficult claiming "our discipline" will be—much harder, or messier, than I would have ever imagined when I was writing about the divide between literature and writing when I was a graduate student.

In short, the struggle to claim the discipline as our own does not automatically emerge by becoming our own department; the baggage, if you will, of being connected to a literature-focused department, remains. In a recent meeting I held with adjunct faculty to hear their concerns, to continue to find ways that we can improve labor conditions for them on this campus, I was astonished to discover that they were not concerned with labor conditions. Their concerns were entirely focused on whether or not they would continue to be able to teach literature in their first-year writing courses. Likewise, many of our instructional specialist faculty have told me directly that they are not going to want to teach writing here if and when the department moves away from a literature-focused writing course.

There are also various frustrations related to service in the department. As Laura explained, when they were part of the English Department, instructional specialist faculty were forbidden to participate in service roles. Now they have the ability to do so, but by contract their service roles cannot factor into their performance reviews, so many IS faculty are opting not to do so. This leads to a division in the department between faculty who serve because they care about the department and wish to be fully involved in the life of the department and those who only focus on their teaching. From my perspective as chair, I am divided about this. On the one hand, I completely understand the desire to only focus on teaching. The IS faculty load is heavy (4/4) and there is no actual financial reward connected to contributing to the service of the department. On the other hand, there is a great deal of curriculum, promotion, and steering work that needs to be done in the department and it cannot all fall on the small number of tenure-track faculty we have. Not to mention the enormous service load of the First Year Writing program and the work that needs to be done by the First Year Writing director.

Caroline noted in one of her memos to me that when the instructional specialist faculty were in the English Department, there was unity because they all shared in the experience of being marginalized by the English Department.

Now that we are our own department, there is division between the instructional specialist faculty, who want to continue teaching literature as they did in the English Department, and the tenure track faculty, who want to move the department in new directions.

One thing that I could not have known until I experienced this is how difficult it is to navigate these tensions when there are actual people involved, people whose teaching I've observed, whose friends and families I have gotten to know. It's not an easy matter to simply say, as I might have imagined earlier in my career, that if people don't want to get on with the business of a new department they can just find employment in an English department somewhere else. As we all know, if that was ever a reality it's no longer the case. And in any case, our current administration seems to have no intention of hiring replacements for faculty that leave. So, I continue to hope that our conversations about the divisions in our department between writing and literature, tenure track and contract faculty, will eventually make our department healthier. After all, collaboration is a central ideal to our field.

At the same time, not all desires are equal. To remain true to the mission of having a writing studies department means that we cannot allow a person's desire to teach literature supersede their desire to teach writing. With that statement, of course, I find myself back in one of the central struggles of our field. And just because that struggle has been documented to the point of exhaustion in our literature does not mean that writing program administrators do not have to continue to deal with it on a daily basis. The ongoing tension between literature and writing even permeates recent arguments about hiring from within our department for the next tenure-track hire, when (and if) that opportunity ever comes.

In our discussions around this chapter, Caroline and I have wondered if, in discouraging first-year writing courses oriented around literature, we are enforcing a kind of normativity in the department that tamps down the kind of interrelationships between fields. Is it possible to "let poetry in" our writing courses as a gesture toward collaboration? I think this issue takes us back to *kairos*. Maybe at some point, but not at this time. As Caroline noted in one of her memos to me, "the work of Michael Warner (2002) reminds us that normativity should not be confused with norms (197). Norms are not inherently bad, as they can provide much-needed structure, as is required in a department and a discipline. Norms are what help us determine whether a set of practices are more suitable to an English department or a Writing Studies department. Norms are what help us define ourselves as a discipline. In applying these norms, we are not implying that the practices that fall outside our discipline are misguided or bad (which is how normativity functions). As departmental leaders, if we don't articulate some norms of Writing Studies (i.e., prioritizing rhetoric over poetics), why call ourselves a Writing Studies department?"

CAROLINE: THE EVOLVING SERVICE LANDSCAPE

I am in my fourth year as the director of First Year Writing, and I have another two years left in my term. Because we decided in our bylaws that the position should be occupied by a person with tenure, it is not clear who might assume the role after me. All of the tenured faculty in our department currently hold administrative roles: department chair; director of Gender, Sexuality, and Women's Studies; and director of First Year Writing. Collectively, these administrative positions leave the tenured faculty less time to engage in committee work at the departmental, college, and university level. What makes this scenario especially challenging for our department is that instructional specialist faculty contracts stipulate that their rehire is conditioned solely on the basis of their teaching. In other words, they are not rewarded in any way for service or scholarship. The tenured faculty remain sensitive to this stipulation, careful (when in leadership roles) not to give the impression that work in these areas is expected. With certain duties being limited to tenured faculty only (i.e., First Year Writing director, promotion review), these faculty will carry heavy administrative and service roles for the foreseeable future.

While we had a vague sense when we split from English that this scenario would be the case, one shift that we did not anticipate is how non-essential committees and initiatives within the new department have slowly melted away, we hypothesize, because of the increased committee demands. Many instructional specialist faculty, though they are not rewarded for it by extrinsic means, have been eager to populate the various committees that make the department function: curriculum committee, assessment committee, executive committee, and a host of others. When we were members of the English Department, instructional specialists and part-time faculty were not allowed to serve on committees per the departmental bylaws. Within the First Year Writing program, however, a robust network of program-oriented committees developed, focusing on intersections with creative writing, readings on campus for FYW students to attend, a program-wide blog oriented around teaching, and in recent years, a podcast. As time has progressed in the new department, these initiatives have diminished or fallen away completely. We believe that increased demands on instructional specialist faculty's time for serving on fewer niche-oriented committees (along with their 4/4 teaching assignment) has led to this curtailing of initiatives.

While these creatively focused ventures have contributed to the vibrancy of the department, we find queer theory helpful here as a way to avoid simply painting these changes as loss. Queerness as *techné* resists binaristic thinking; it is rarely productive to see departmental changes within a strict good/bad framework. We find that often when faculty frame new developments as bad, they are tapping into a nostalgia for a time that, though not perfect, was more comfortable or recognizable in some way. For example, it might have been more comfortable work to sit on a niche committee that did not carry much influence rather than serve on the curriculum committee, which engages with fundamental questions about what the department is doing and why. As Ron wrote to me in one of his reflective memos, "I think we are finding some self definition, even as we branch out to providing further services to other departments. I think if we can move to creating a writing course that does not work from deficit assumptions but sees writing as a positive, content-generating discipline, we'll thrive." Even though the service-related landscape is drastically changing right now, it will benefit us to maintain a focus on the core of our mission for teaching writing. And in some ways, our work as teachers of first-year writing may be invigorated by a more streamlined attention within the department to that two-course sequence (and how it can form a foundation for our major). Queerness as *techné* helps us consider how to talk about these changes to our departmental colleagues and upper administration as well as be nimble in charting our future plans.

As administrators, we also do not want to be directive in suggesting which committees are more important than others for the health of the department. While the department cannot function without some of them (i.e., curriculum), others that are more creatively focused bring a character and sense of community to the department. In writing about the changing nature of service in our department, Ron asked me if we as administrators are being normative as we try to ensure that the department remains focused on essential business; in doing so, are we implying that more creatively focused initiatives are not important? While we don't have easy answers at this juncture, queerness as techné may provide guidance for us as we try to avoid being normative about what a department should look like. Just as techné celebrates both theoretical knowledge and practical skill, we seek to avoid a hierarchy of more/less important service work within the department. Working collaboratively among all ranks and positions, the department will shift its focus over time depending on changing circumstances and do so as a group.

RON: THE WRITING STUDIES MAJOR AND ITS MISSION

The tension that Caroline describes between a department's theoretical grounding and its service expectations plays out in ways that go beyond interdepartmental relations as well. Even though scholars in composition and rhetoric have been producing knowledge independent of service for more than fifty years now, there remains a tendency, including among our upper administration, to conflate writing studies with first-year writing. This increases the difficulty of claiming a discipline within a university. This reality of this tension was first introduced to me when our college president decided to delay our major in public and professional writing.

Toward the end of my first semester of being department chair, I was given the news that the president of the university had read our Program Announcement Proposal for our major and was going to block its moving forward. That semester the department had already pushed the major through the college and university curriculum committees, and I had incorrectly assumed that the Board of Trustees meeting was only going to be a formality.

The president's concerns, as delivered to me by the provost, were that there were too many remnants of our connection with the English Department in our current course offerings and that there were not enough efforts in our new major for us to collaborate with other colleges. Although I did not fully agree that our courses sounded like they should be in an English department, it seemed wise (or strategic) to take the opportunity to revise some course titles and descriptions, and even take the opportunity to create some new courses. (To my surprise, these course revisions were immediately put through by the president without having to go through any of the required faculty committees.) In the big picture, this part of the process was relatively painless.

What was more disconcerting was the imperative to collaborate with other colleges and schools across the campus. The provost told me that they had just spent five million dollars on new equipment in the School of Communication and that we needed to collaborate with them. I explained that our courses in technologies of writing, digital and technical writing did not require five-million-dollar equipment and that we had quite a few departments that we were already forging meaningful collaborations with, but it turned out that this particular collaboration was a deal breaker.

Seeing us primarily as a service department, the upper administration was, they told me directly, more than happy letting us be a department that provided only first-year writing and a minor. I could not bear the thought of letting the major proposal die in a file somewhere, so I set out on the long, rather tedious process of scheduling meetings with the dean and faculty of people from the School of Communication, a group that had no organic interest in working with us, and who—understandably, I'd say—resisted the imperative from above to "collaborate." And, truthfully, there is not a lot of story to tell. I waited outside of offices for scheduled meetings where no one came. I met with people, but when it comes down to it, I never got anyone to agree to anything, but I did explain the situation to people when I met with them. Understandably, both the dean and the faculty I met with explained that they did not want to move forward without further faculty input from various departments within the school.

One day, though, I received an email that our revised major would be approved at an upcoming Board of Trustees meeting. And so, our revised major was born, a year later than we had expected and, as far as I am able to tell, without additional signatures from anyone.

***

With all the circumstances surrounding it, our major's finally coming into being felt like a success and a failure all at once. It set the groundwork for a possible future, but in a way that made it clear we would have to be particularly strategic about what the "public" in our major in public and professional writing actually meant, at least at the level of what is seen from the outside. As Michael Warner (2002) puts it,

> The critical discourse of the public corresponds as sovereign to the superintending power of the state. So the dimensions of language singled out in the ideology of rational-critical discussion acquire prestige and power. Publics more overtly oriented in their self-understandings to the poetic-expressive dimensions of language, including artistic publics and many counterpublics, lack the power to transpose themselves to the generality of the state. Along the entire chain of equations in the public sphere—from local acts of reading or scenes of speech to a general horizon of public opinion and its critical opposition to state power—the pragmatics of public discourse must be blocked from view. (116)

The imperative to remove all vestiges of the English Department from our program had more to do with making sure that our president could market this degree as one that was focused on practicalities than it did with being interdisciplinary. I was asked to remove a great deal from our public announcement of the major, including references for our desire to correct inequities related to race, gender, and sexual orientation. I was asked to reduce the core courses in our major to thirty credits plus twelve from other colleges within our university. What resulted was a major that was interdependent on what our college president believes to be more marketable colleges (business and communication, primarily). At the same time, though, we were able to shape our curriculum in some interesting and compelling ways in the process. As Caroline said to me in a memo about this process, "My first reaction is to shake my head at what you had to go through in the process of getting the major passed (which happened largely when I was on sabbatical). In truth, I think that [our president and provost] were on the right track in saying that our initial plan was still too tethered to English, but the process by which they let that be known is quite appalling. And then the remedy of forced collaboration is not one that helps foster long-term trust between disciplines."

In this regard, we find our situation very much in conversation with Duffey in this volume, when she writes the following:

> To understand what disciplinarity is, where it resides, how it plays out in the lives of scholars, practitioners, and other professionals who claim it, and how institutional frameworks intersect with disciplinary identity, intersections that have less to do with knowledge-making and other professional, pedagogical, and scholarly goals and negotiations than with bureaucratic and political ones.

The university pushed us in the direction of practicality with the cognate areas designation, and in our case the negotiation with bureaucracy had profound implications on what our discipline looks like to our students. Through one lens, this particular anecdote illustrates how writing studies as a discipline continues to be marginalized, but now from a more distributed place. Just as many in English departments do not respect writing studies as its own field, it was clear that our upper administration felt that a major in writing studies could be redefined through their own views of what was needed at the time.

At the same time that I still wrestle with the ramifications of this failure, I now see my choice to be strategically passive as one that has put my department in a position to more fully reach students, and in this regard, it turned out to be an effective "pivotal strategy." As Caroline has put it, "viewing queerness as *techné* helps us to reorient toward the process of adaptation, the flexibility of method, the need to constantly change our approaches." This cognate area idea was not our original intention, but we decided to remain open to its possibilities in the hope that it would connect us more meaningfully with students. At least right now, it appears that students are more attracted to courses like Workplace Writing and Grant Writing than more overtly political courses, such as Race and Rhetoric and The Rhetoric of Political Writing, as the latter courses have had to be closed recently as a result of the lack of interest.

I think overall we are still in a learning stage when it comes to what it means to claim our own discipline. I think it depends entirely on who the "our" is in "our own." Putting administrative pressures aside, how should a department respond to its students' desire for practical courses, at least at the level of the title? At the level of curriculum, should issues such as race, class, and gender be distributed across a series of practical courses rather than be offered as independent subjects themselves? These are questions that future students in rhetoric and composition should be given the opportunity to flesh out more fully before arriving in leadership positions at universities.

From my perspective, rhetoric as a subject matter is flexible enough to remain connected to its humanistic mission as long as we remain attuned to its flexibility. On balance, though, I'd say wrestling with these questions within our own department is preferable because we are now in a position to move forward and to grow, if not toward independence then toward a model of interdependence that will work its way toward, I hope, a basis in mutual trust and sincere need.

CAROLINE: CURRICULAR REVISION

As Ron mentioned earlier in this chapter, the First Year Writing program is undergoing a revision of its two-semester sequence of courses. The second-semester course is being revised as a result of an institution-wide remaking of the general education curriculum; Caroline initiated a revision of the first-semester course in order to align it more directly with first-year writing curricula from around the country. This process has caused us to think about how our professional identities have influenced and been influenced by the subsequent departmental discussions. While not applicable to independent departments in general, the structure of our particular department illustrates how much the labor model of our discipline (where we rely mostly on contingent faculty to teach first-year writing) is an always-present factor in our disciplinary work. Our instructional specialist faculty are not required nor rewarded for engaging in any kind of scholastic activity, and therefore we do not see sizable numbers of faculty attending conferences or publishing in the field. We believe another limitation to disciplinary growth is the contractual stipulation that instructional specialist faculty cannot be paid overload, including stipends for attending professional development sessions that the department sponsors. As a result, we have seen attendance at professional development sessions steadily decrease since the introduction of the instructional specialist contract. When it comes to revising curriculum and integrating the latest pedagogical approaches to teaching first-year writing, then, how can we enact a collaborative model with so many faculty who have been contractually limited from staying current in the field? How can department leaders maintain a fidelity to disciplinary commonplaces while also trying to honor the voices of contingent faculty?

For us, this scenario encapsulates some of the messiness that attends an independent department that is comprised of majority contingent faculty. As more writing studies programs split off from English, most of them will likely have large numbers of contingent faculty who will journey to the new department. Their ratio of tenured to non-tenured faculty may not resemble ours, but the very act of splitting off means that a new department will have to consider how to handle the labor dynamics in a way that they did not have to while in English (where hierarchies are, in general, more entrenched). In trying to be a First Year Writing director who is shaping curriculum in a way that is broadly accepted in the discipline, Caroline finds herself in moments of needing to "educate" her colleagues who do not have the privilege of contractual release time for research about current trends in the field. In a memo reflecting on this scenario, Ron wrote, "I wonder if being an 'educator' in this way makes us normative?" This concern is often present for Caroline who, as a queer writing program director, tries to enact queer methodological practices in her administrative approach. Attempting to embrace the messiness of curricular revision in our particular context has meant enacting a less nonlinear approach than had been used in the past, with pilots of new curricular models co-existing with courses taught according to the current curriculum, and a timeline for revision that is malleable and adjustable depending on what transpires at department and first-year writing committee meetings. Caroline has tried to follow Ron's model of administrative leadership in scenarios such as this; a model that relies less on the educator role and more on the consensus model. As she wrote to Ron in a memo, "I appreciate that your inclination as Chair is to give people time, let them work through their anxieties, talk it through—and then return to the question at hand. Shared decision-making might take longer in our department, but ultimately, I can see that this approach will make the department stronger." We also believe that our particular departmental landscape illustrates a conflict between some of the field's idealism about labor and the reality that we are encountering. Using the frame that Bowen and Pikert set up in chapter 1 of this collection, we have found in this context especially that "political and economic circumstances" often outweigh "personal circumstances" and even "pedagogical imperatives," which can often come from more idealistic space. The big question for us remains: When contingent faculty are contractually limited from engaging in scholarship and service, and they cannot be paid to participate in professional development, how can the discipline continue to develop and thrive in an ethical manner?

Conclusions

As we come to the current place we are with our department, both of us find it difficult to find conclusions. We agree, though, that the following observations have emerged:

1. Starting with the ideal of inclusiveness, we wanted to create a department that treated all of its members that teach first-year writing, if not equally, as equally as conditions allow. This openness has made self-definition more difficult, as the process of listening and responding to resistance from within our department creates a great deal more labor. We do still believe, however, that our collaborative process is one worth pursuing. And we think it is a model that should be taught more to graduate students with a specific focus on some of the difficulties of embodying such an approach.

2. Along similar lines, it has been worth listening to our students as they help shape the direction of our major. Many English departments have been slow to respond to market realities due to a desire to keep a degree of disciplinary purity. At the same time that we do not want to let "the market" dictate our curriculum, we do believe it is important to listen to our students. As a majority-minority institution and a Hispanic-Serving Institution, Montclair State has a student population that requires targeted advising and timely completion toward degrees. Our students' desires for practical courses are meaningful. At the same time,

we have a responsibility to keep writing studies connected to principles of social justice that go beyond marketability alone.

3. In imagining more equitable labor models, queer methodological concepts such as failure and messiness have helped us establish nimble approaches for how to move forward. Not recreating the kind of hierarchies that we saw within the English Department is an important goal for our nascent department, one that is complicated by the fact that our profession encourages an "us versus them" mindset when it comes to contingent faculty. We recognize that the makeup of our department requires that we use creativity when meeting challenges. Our solutions often include messiness, and we experience frequent failures. We view these experiences as instructive for how we can be strategic during future challenges. For faculty at other institutions in a similar position, we recommend that they focus on the potentialities of an independent department; how things might be different. Constraints will always emerge, but having a clear sense of the core values of a new department will help spur the creativity needed to make progress.

Notes

1. As part of the separation agreement between the English Department and the Department of Writing Studies, both parties agreed that all Writing Studies graduate courses would be sunset by a certain date. Doing so would allow Writing Studies to begin its own graduate program in due course, and the English Department could not continue offering Writing Studies graduate courses, as was their stated interest.
2. The dean took these two proposed lines, which were rejected in the English Department, and used them to hire two faculty for the new Department of Writing Studies.
3. Notably, the approximately fifty adjunct faculty in the department at that time were not invited to the photo session.

References

Ahmed, Sara. 2006. *Queer Phenomenology: Orientations, Objects, Others*. Durham, NC: Duke University Press.

Dadas, Caroline. 2016. "Messy Methods: Queer Methodological Approaches to Researching Social Media." *Computers and Composition* 40 (June): 60–72. https://doi.org/10.1016/j.compcom.2016.03.007.

Everett, Justin, and Cristina Hanganu-Bresch. 2017. *A Minefield of Dreams: Triumphs and Travails of Independent Writing Programs*. Fort Collins, CO: The WAC Clearinghouse.

Halberstam, Jack. 2011. *The Queer Art of Failure*. Durham, NC: Duke University Press.

Muños, José Esteban. 2009. *Cruising Utopia: The Then and There of Queer Futurity*. New York: New York University Press.

O'Neill, Peggy, Angela Crow, and Larry Burton. 2002. *Field of Dreams: Independent Writing Programs and the Future of Writing Studies*. Logan: Utah State University Press.

Warner, Michael. 2002. *Publics and Counterpublics*. New York: Zone Books.

SECTION II

Negotiations and Resilience

INTERLUDE TWO

Negotiations and Resilience

Like the literacy narrative assignments instructors may require of first-year students, stories of claiming (and being claimed by) writing studies may lead readers to expect epiphanies. These are predictable stories whose narrative arc follows the well-trodden path: "I was lost . . . until I found" and emphasizes the absolute happy ending. In fact, epiphany narratives, like hero narratives, reduce the real complications of human experiences to simplistic form. The narratives in this collection, and especially in this section, instead detail the stone-by-stone building of self and collective identity. In this next section, Ersheid, Konigsberg, McVeigh, Pearson and Kahn; Johnson; and Tellez-Trujillo negotiate experiences redolent with pain. Ersheid, Konigsberg, McVeigh, and Pearson are trained creative writers who have stumbled or slid into writing studies and find themselves both claiming and being claimed by the field. Their co-writer, Seth Kahn, provides the institutional contexts and histories necessary to their narrative as well as a keen awareness of their emotional labor. These collaborators live precarity, to use Johnson's term, which conditions their lives as teacher-scholars. Johnson, likewise, describes her stance as queer, working-class academic and deliberately eschewing epiphany, finds resilience in "tangled positionality" (this volume) and foregrounds how precarity circumscribes the availability of choices as does Tellez-Trujillo in her experiences as a bilingual Latinx woman. Tellez-Trujillo's and Johnson's bodies and identities clash against the expectations and norms of their teachers and professors—their emotional labor along with the others in this section suggests that stubborn effort; deep reflection; and refusal of the names, identities, and vocabularies that powerful actors insist on constitute key components of these teacher-scholars' pivotal strategies in claiming writing studies.

Time is not on their side although moments of kairos dot the landscape. There is not enough time to embrace all parts of these contributors' writerly identities, creative writer and writing teacher, queer scholar and working-class teacher, or Latinx woman and rhetoric graduate student. Becoming is the goal, perhaps. Becoming what is the endless open question.

To note here: the act of writing out these experiences is emotional and requires strength of mind—resilience—for sure. It is itself strategic. The telling, in other words, sanctions the experiences. While not epiphany narratives by any means, the metacognitive affective work necessary to write about personal-professional identity must be understood as strategic—and finally, as necessary to making the claim for writing studies.

5
Claiming and Being Claimed by Writing Studies

Negotiating Identities for Creative Writers Teaching Composition

ALISON ERSHEID, LISA KONIGSBERG, MAUREEN MCVEIGH, NANCY PEARSON, AND SETH KAHN

Creative writers (identified by degree, publication record, or both) teach composition courses extensively in the US. Since the mid-1980s, compositionists and creative writers have written frequently about curricular and pedagogical tensions between the two and made attempts to synthesize the fields (including Hesse 2010; Bishop 1993; Mayers 2005; and many others) in philosophical and practical terms; our chapter approaches the problem less as a matter of philosophical differences between creative writers and compositionists and more as a *labor problem*. We present a *collaborative autoethnography* (Chang, Ngunjiri, and Hernandez 2012) of a department where labor conditions for writing studies specialists and for non-tenure-track (NTT) faculty are favorable, but where those conditions have demanded difficult negotiations for composition faculty who identify, personally if not professionally, as creative writers: negotiations between *claiming* and *being claimed by* writing studies.

Our story is set in a department and institution informed by complex policy changes and politics, and the way we address those is akin—if not identical—to Michelle LaFrance's (2019) institutional ethnography (as juxtaposed from Brooks and Dadas's queer collaborative autoethnography in chapter 4 of this volume). It also functions, in a way we didn't anticipate, as a sort of "hidden transcript" (Scott 1990) responding to the kind of exchanges Tara

https://doi.org/10.7330/9781646426331.c005

Wood describes in chapter 3; while she describes complex negotiations over disciplinary status among administration, faculty across the institution, and herself as a disciplinary expert, our chapter presents the perspectives of writing instructors whose careers hung on the other end of those decisions and who needed to think about how to position themselves in conversations to which they were not directly invited.

The story starts in 2007, when changes in our union contract altered the consequences of hiring off the tenure track by enabling the possibility of converting long-term non-tenure-track faculty onto the tenure track, and right before Lisa (in 2008) and Maureen (in 2009) were hired; we'll narrate the tenured/tenure-track (TT) faculty's negotiations of policies to respond to those changes. We'll describe the impacts of those policies personally and departmentally, how they set the scene for Nancy and Alison's hiring in 2017, and their experiences since. We'll end with recommendations for departments that hire faculty whose identities are similarly complex. In a nutshell, we call for departments to *help creative writers claim writing studies, but not to coerce them into it*. Everybody but Seth (who started in the field, so never pivoted from anywhere else) had to navigate the moment of pivoting at more advanced stages of our careers. Therefore, much of the story is about the difficulties of making that move, including decisions about how far to pivot from our original identities, under at least tacit pressure to do so.

Introducing Ourselves

Of the five of us, only one (Seth) identified as a writing studies specialist when we came to West Chester. Given our focus on the push-pull of claiming/being claimed into the field, this section introduces us with an eye toward establishing what we did claim, before the labor situation here called on us to pivot—for Alison, Lisa, Maureen, and Nancy, pivoting into writing studies for the sake of job security and having to decide how far the labor situation required them to move; for Seth, pivoting to a new way of understanding membership in the field that wasn't entirely grounded in graduate course work and research training.

ALISON: My path took many meandering turns before landing me in academia, though my focus on writing never wavered. I graduated with a BA in English and a minor in writing from a small, satellite state university and, since the tech boom was literally reshaping life and industry as we knew it, technical writing jobs abounded and paid well. Technical writing led to marketing writing, which led to project management and, soon after, I was looking for the door. Writing, which had begun as a deep and meaningful pursuit, had evolved for me into a dry, lifeless, transactional way to make a living. Corporate work and cubical living, I had discovered, were not intended for me.

Graduate school had never crossed my mind before this point. I came from a modest family and community that regarded a four-year degree as bonus points, so anything beyond that never occurred to me. I pendulum-swung into the highly regarded MFA program at George Mason University and, as I practiced pouring out all those years of pent-up creativity, I became a TA for the university. At first, I considered the tuition waiver and stipend to be the prize, but, after one semester in front of the classroom, the discovery of my love for teaching overshadowed all else. I taught, and was trained to teach freshman composition courses, but I figured one day that would turn into teaching whatever struck me. Again, I had absolutely no concept of the machinations of academia and the complex processes and politics involved in higher education employment. Upon graduating in 2007, I worked for many years as an adjunct, teaching mainly first-year writing courses with a few creative fiction classes mixed in. It became clear that composition courses were much more readily available and so I began to focus more exclusively on marketing myself to be more attractive for those jobs. That meant spending less time writing creatively and more time trying to mimic the moves that would better position me as a capable writing studies specialist.

LISA: I come from a working-class family whose first concern was always having employment, while simultaneously respecting education and scholarship. So, adjunct teaching was something I was prepared to do to remain in the field. I taught at three different colleges for over ten years before I was offered a full-time course load at West Chester. At the same time, I neglected my creative writing, squeezing poetry and bits of two novels (still in progress) between teaching and grading. Because revising work takes focus, it has been difficult to find that kind of space for my creative writing. Though my passion has always been poetry, I worried that an MFA in creative writing would not lead to a full-time position, so I abandoned that idea in favor of an MA in English. That decision seemed to make sense at the time (the 1990s) because jobs in CRW were hard to find. To comply with WCU's 11-G conversion policy (which we explain below), I completed twelve credits in the Composition and Rhetoric graduate program at Temple University between 2016 and 2018.

My sense of belonging in the field of writing studies came from thousands of hours in the classroom teaching mainly first-year writing. Pivoting between composition and rhetoric research/teaching and my own writing has been the most frustrating aspect of my professional life. When I was awarded tenure at West Chester University in 2019, I was relieved.

MAUREEN: I took a circuitous path to this career after a BA in international affairs and economics and working for several years before beginning an MA in English (creative writing) at WCU. After graduating in 2005, I began an MFA–fiction program while teaching composition at local community colleges. I felt unprepared to teach those courses but recognized how important they were to students both as academic requirements and as preparation for other courses. I spent more time researching FYW pedagogy than I had anything in graduate school. Some schools gave me a required textbook and syllabus. Others allowed more freedom. In order to be an effective instructor, I read as many FYW textbooks and contemporary pedagogy journals as possible. For each assignment and objective, I found examples from other instructors and institutions to inform my own. I constantly updated materials based on student feedback and success. A silver lining to teaching at so many different colleges was that I taught an extremely varied student population. In addition, the diversity of community college students was incredible teacher training. Any doubts I had about teaching FYW instead of CRW were overruled by the satisfaction of helping nontraditional students more fully enter the academy and watching students improve their critical thinking. I was claimed by writing studies in order to get into teaching, but I fully claimed it for myself once I began.

NANCY: I started teaching composition in a maximum-security prison near DC. With only a BA in English, I was highly unprepared to teach anything, much less teach writing to students who mostly spoke Spanish. It was there, however, that I realized my experience reading and writing poetry (primarily self-taught) helped me connect to my students and helped my students make sense of their world. This experience inspired me to get an MFA in poetry and later an MFA in creative nonfiction. Like Maureen and Lisa and many others with creative writing degrees, I primarily taught first-year writing in community colleges hoping my experience teaching as an adjunct would lead to a full-time position. Despite my search, I ended up teaching in adjunct positions for over fifteen years to support my family and continue my teaching dream. During this time, I never considered myself a writing studies specialist. I sensed that my CRW degrees were handicaps, obscuring my identity as a "pretty good" first-year writing instructor. I never thought my achievements, including two award-winning poetry books, would advance my career as a writing instructor. But I continued my scholarly pursuits in poetry with the belief that scholarship—even the kind that's not recognized in a writing studies field—would contribute to my self-worth, as both a writer and a teacher. I wrote prolifically, presented at conferences, and participated in readings nationwide. I identified personally and professionally as a poet foremost. I never claimed writing studies, although I was doing the work.

SETH: My story is closer to those Bowen and Pinkert tell in chapter 1 than to my co-authors. I wandered into a job tutoring composition at a community college in 1994 and loved it, which is how I discovered the field of composition and rhetoric, having graduated with a BA in history and philosophy. My MA from a department (English at Florida State) where rhetoric/composition was nearly autonomous and my PhD from a freestanding writing program at Syracuse, made me narrowly focused on writing studies from the beginning; for instance, I didn't pivot from another field into this one. What pivoted was my sense of how inclusive our professional identity is and should be. That transition happened as I came to understand how many contingent faculty are teaching writing successfully under deplorable conditions, and how much better that work could be for them—and for those who struggle to do it well for lack of training—if the working conditions and support/professional development were better.

Institutional/Departmental Context

WCU is a large (approximately 17,000 students) public regional-comprehensive university in the Pennsylvania State System of Higher Education. All faculty (NTT and TT) on all fourteen campuses are represented by the same faculty union, the Association of Pennsylvania State College and University Faculties (APSCUF), and work under the same collective bargaining agreement, which has garnered attention for its protections against the worst abuses of contingency (see Bousquet, Hendricks, and Mahoney 2009). The contract has earned praise for decent pay and access to insurance and retirement benefits; grievance rights and academic freedom protections; and shared governance provisions, including NTT faculty voting rights. We struck successfully in October

2016 in large part to reject a proposal that would increase full-time NTT teaching loads from 4/4 to 5/5 with no increase in pay.

Our department, English, houses about twenty (depending how you count, which is part of our story) writing studies specialists among approximately seventy NTT and TT faculty, representing a vast array of specializations: literary periods/genres/theoretical approaches, linguistics, film studies, creative writing, professional and technical writing, comparative literature, digital humanities, and more. The practice of assigning general-education composition courses to almost everyone regardless of specialty, and rarely assigning anything other than composition, business, or technical writing to NTT faculty, has located composition in a messy juxtaposition of workload and (sub)disciplinary identities: composition as both a specialization (i.e., *claiming* writing studies) and a burden we must all share (i.e., *being claimed by* writing studies). The most unusual feature of our department is the size of our writing studies cadre, one result of which is that writing studies specialists aren't an oppressed minority in the department; we (some of us) teach throughout the curriculum and are well-represented in departmental leadership, including four of the last seven chairpersons. Other than the fact that both institutions are unionized, our situation couldn't be much different from the situation at Montclair described in chapter 4. While there are occasional conversations over beer or coffee about forming a writing studies department, none of those has reached the ears of a dean since the late 1990s.

As a feature of institutional and departmental context, it's important to note that WCU and the English Department struggle with racial diversity, and it's not lost on us that all five of us are white. That's not to say that the NTT positions of everyone but Seth weren't precarious, but the focus on race in so many chapters in this book makes us recognize that many of the issues we face are at least less fraught than those of BIPOC colleagues in similar positions.

2007: Changes to the Union Contract Set the Scene

In 2007, three changes to our union contract helped set the scene we're working in. The first change removed a provision called the McGuire Memorandum, requiring NTT faculty to take every fifth semester off; we called it "getting McGuired." That change activated a contract provision, Article 11G, which we'll refer to frequently, allowing NTT faculty to convert into tenure-line positions via department faculty majority vote after ten consecutive semesters of full-time service. The other change was a new 25 percent cap on NTT faculty on each campus. Our university exceeded the cap for years, which eventually created pressure to hire TT faculty and encouraged management to support converting NTT faculty more than they might have otherwise.[1]

2008–2009: *The Old Normal*

The effects from those changes to the contract took a few years to become visible. In the meantime, like many English departments, we hired adjunct faculty, sometimes more systematically, sometimes less so; we hired people without terminal degrees or composition/rhetoric graduate course work into what were ostensibly "temporary" (the term our contract uses for NTT) positions as long as they had teaching experience, which would eventually lead to conversations about the status of writing instructors and instruction much like those that Brooks et al. (chapter 4) and Wood (chapter 3) recount. While management resisted multi-year contracts because they would require seniority raises, nobody objected when NTT faculty who wanted full-time schedules got them.

LISA: I was hired in fall 2008 as an adjunct on a semester-to-semester contract and continued that way until I got converted (we don't have renewable positions like Laura Field describes in chapter 4) usually given a full course load. I taught at several colleges while teaching at WCU and had no knowledge of 11G prior to taking the position. Since WCU offered higher pay and a full course load with benefits, it was the obvious choice for someone who had spent fifteen years as a contingent faculty member trying to work in the field I felt I had been preparing for and where I could apply my skills as a teacher and writer, however precarious the possibility of permanent employment was. I wasn't clear about my role in terms of how I fit into a scholar model or where my place was in the field of comp/rhet instruction. I simply wanted to teach writing on the college level.

MAUREEN: I was hired at WCU last minute to fill in for a creative writing faculty member on leave in fall 2009. The following semester, I was offered a full-time schedule of composition courses. By then, I had taught composition, creative writing, and children's literature at seven schools over several years. By 2009 I had created syllabi, lessons, and assignments that worked for students and aligned with FYW goals. For the first few years at WCU, I still taught at other schools due to the precariousness of the adjunct situation.

After several consecutive years of full-time course loads, I taught only at WCU, both composition and creative writing, including graduate creative writing courses, and even advised creative writing MA thesis projects. I had no knowledge of 11G, but WCU offered some stability, higher pay, union membership, and more diverse courses. In addition, the English Department offered both flexibility and support in FYW pedagogy.

SETH: The only conversations I remember about NTT labor during this time expressed relief that faculty wouldn't get McGuired anymore.

2010–2012: *Debating Conversions and Making a Policy*

The timeline here is a bit complicated. Article 11G, the conversion article, had been in the contract but moot for many years because the McGuire Memorandum kept anyone from reaching ten consecutive full-time semesters. When McGuire was removed from the contract in 2007, the result was that anyone hired in 2006 would become eligible for the conversion vote at the end of the 2010–11 school year (management and the union had agreed that departments could conduct the votes during the tenth semester instead of making the candidates wait an extra year). The vote is mandatory (i.e., if a faculty member doesn't want to convert, we still have to vote—the faculty member just tells us to vote no), but ambiguous contract language only requires departments to have "a process" for conducting it. That ambiguity created—perhaps exposed—tensions we'll describe below.

SETH: Our first department meeting in fall 2010 suggested how complicated things would get. Sticking points were soon clear: the value of national searches versus converting long-time colleagues; and the importance of terminal degrees focused in writing studies. Like many departments, we had been hiring faculty with MAs in English, like Lisa and many others, and faculty with creative writing MFAs (like Alison, Nancy, and Maureen), to teach first-year writing.

LISA: I was happy to teach composition courses at WCU as the larger part of my teaching career was seated firmly in that subject matter, which meant I felt capable of teaching comp effectively. (It's important to note that feeling confident isn't often part of the adjunct's experience.) Although I hold a master's degree in English literature, I had spent the greater part of my twenty-year career teaching composition. I was given composition textbooks to teach with and had to work from them, often verbatim. I also had already seriously considered pursuing a PhD in composition and rhetoric but did not have the time or money.

MAUREEN: When the creative writing faculty member I filled in for returned, the department chair offered me composition courses, which I happily accepted. I appreciated the department's commitment to composition and support for faculty. Due partly to my qualifications and to department need, I continued to teach a variety of CRW courses as well. At that time, I felt very connected to the CRW students and their success as writers. A group asked me to advise their independent writing group, a student included me in the dedications of her novel, and an alumnus hired me to teach in the MFA program he ran. I felt little tension teaching in both disciplines as I always had, and every CRW faculty member at WCU also taught FYW. I began to compile research to support my pedagogical decisions to transfer strategies from one discipline to the other. For example, peer workshop is foundational in CRW and I found it as helpful in FYW. WCU requires faculty observations, and I received positive feedback from my peers on using workshops in FYW.

SETH: In my twenty years at WCU, this debate is the fiercest we've had. The concrete issues related to hiring ethics spun into arguments about future hiring, which spun into arguments about writing studies/literature tensions that most of us thought long resolved. We had formal meetings; subgroups organized meetings; people talked in the hallways and at bars and conferences. Friends raised their voices at each other. It felt like everybody was buzzing about it, but, honestly, those of us invested in either side had no idea what most of our colleagues were thinking or what the potential convertees were feeling. Plus, we had to make a decision before the first potential candidate hit the eligibility threshold.

MAUREEN: One of the things I like about the WCU English Department is its commitment to social justice and union principles. The debates supported this and created complexities regarding 11G in surprising ways. Since I had been an adjunct for years, I was used to feeling like a second-class citizen, but the prospect of 11G made me cautiously optimistic. Initially, though, my primary feeling was confusion. I did not understand the expectations to apply and wondered if the fierce resistance to it from some faculty was reason enough to avoid it.

In information sessions and conversations, it quickly became apparent that in order to be considered for TT conversion, we would need to show a record of service and scholarship. This, however, was not codified anywhere because our semester-long contracts only included teaching. In essence, it would have been unpaid labor, but those friendly to our cause let us know we had to show the work we would continue to do if we wanted the department to consider us for tenure track.

SETH: We eventually compromised: tenure-track status was OK, but teaching assignments wouldn't change much. Faculty who taught composition or business writing could have security and better pay as long as they didn't compete for "better" courses. Further, because they would be teaching primarily composition and business writing, their scholarship needed to (re)focus on writing studies, whatever their training or professional interests, an obvious source of the tension between *claiming* and *being claimed*. We later revised our department's NTT hiring policy to privilege terminal degrees and national searches for full-time positions (Alison and Nancy were hired under this policy) and formalized an expectation that faculty with literature-focused MAs would do doctoral coursework in writing studies or related fields (e.g., literacy studies).

Put another way: we extended the metaphor of *conversion* beyond a change in status, so that faculty who teach mostly composition would be identified (*claimed by*), and would work to identify themselves (*claim*), as writing studies specialists.

LISA: To "claim" writing studies, I essentially accepted my role as a teacher-on-call, while simultaneously recognizing that I was lucky to be working in the field in a semi-stable position. It was understood among my peers in the field that if we wanted to teach in an English department, we would need to narrow our view about ourselves as teachers, focusing on the needs of the market. Regardless, I embraced the teacher that had always resided in me. I also knew that I had to separate the creative parts of myself. The teacher in me was sitting firmly in my soul next to the poet who always felt left out of the conversation. For the next fifteen to twenty years I continued to try to appease the poet while nurturing my teacher-self by writing poetry on the side. This split between my creative writer self and teacher self widened as I was forced (to some degree) to spend more time teaching courses in business and organizational writing in order to keep my job.

2013–2017: "HOPE IS THE THING WITH FEATHERS" (EMILY DICKINSON)

Our first positive conversion vote happened in spring 2012, and the faculty member started in their tenure-line position in fall 2013. We successfully converted ten faculty during this period, and it seemed like all was well.

SETH: The tensions didn't disappear once our policy was in place. But when we converted our first NTT faculty member, people who had fought for the conversions felt like we'd accomplished something important. From 2013 to 2017, we converted ten faculty. The fact that four convertees had terminal degrees (two MFAs, two PhDs) probably helped dissolve some resistance; so did ten more faculty doing service without having to worry about the labor ethics (i.e., compensation) of NTT committee work (see Brooks et al. in chapter 4 of this volume). Our first two promotion candidates (the two with PhDs) were successful in 2017. The contract and policy were working.

NANCY: Serendipitously, in 2014 I taught a group of WCU creative writing students at an emerging writer's conference at the University of Houston. In fact, all but a few students in my creative nonfiction workshop were students at West Chester. These students were lively and hardworking writers, reflecting the dedication and expertise of the creative writing faculty, including Maureen, before I even met them. Even then, years before I was hired, I wanted to be part of what was happening at WCU, to join a dedicated department where faculty cared enough to bring an entire pack of students all the way to Houston to work on their creative writing manuscripts.

LISA: My conversion to TT through 11G in October 2014 was a strange semi-miracle, because there was an undercurrent of devaluing of adjuncts that I had been experiencing personally for over a decade. When conversion to tenure track should have been celebratory, what I experienced was intense relief. The need to prove myself beyond the observations and student evaluations that occur every semester was finally eased up a little and I could imagine some job security. Conversely, as an adjunct I had already become a member of the committees I had been appointed to by my department chair, but the shift in my responsibilities to the university felt suddenly more concrete and valuable. My view of teaching had not changed, nor had my willingness to support my peers whenever possible. What changed significantly was a sense that I would have to prove I was worthy of being granted a tenure-track position.

Prior to working at West Chester, I had taught at various colleges in the area and had no real stake in the outcome for those schools and their students. What 11G provided was a chance to become a stakeholder. Among my other roles on five committees, I am currently on the Student Retention and Recruitment Committee, which allows me to collaborate and conduct research with faculty across the university. It's not lip service to say that I feel a sense of pride and responsibility outside of the classroom now too, which I could not claim before. At the same time, I was teaching a full course load, and I continued to write poetry in my free time. My first collection of poems, *Invisible Histories* (Spruce Alley Press) was published in 2015 (and another collection, *The Golden Mean* [Moonstone Press] was just published in August of 2019). The publication of my poetry after twenty years in the field of teaching and at sixty years of age was a watershed moment in my life. The books embody the imaginative and aspirational aspects of my creative side that could only be expressed this way. Job security also grants time by virtue of the sense of relief it offers. Adjuncts rarely have time to write creatively. Theirs is a cog-in-a-wheel existence coupled with a sense that they are on a string, liable to be cut loose at any moment. No one works well under those conditions.

MAUREEN: I still recall where I was when the department chair called in October 2014 to say that I had been converted from NTT to TT through 11G: in the classroom. I'd stayed after class to work and that mundane moment transformed into one of the most important of my career. I want to echo Seth's point about accomplishing "a really important thing" as I recall sending him and our then chair a note of thanks for all of their hard work that led to this opportunity for me. When I told friends or other adjuncts I met at conferences about 11G, they were astounded and often jealous. Applying for conversion had been complex, rewarding, difficult, inspiring, and emotional. Lisa and I had spent years in the "adjunct office," which held six unassigned desks shared by at least as many faculty, some of whom changed every semester. Sharing this space allowed us to see colleagues go through the 11G process before us, which was invaluable. At the same time, tenured colleagues who did not agree with the conversion process made this clear individually and in group settings. I did receive support from many tenured colleagues and my fellow adjuncts, though. Compiling the application materials served as a good reminder of all I had accomplished and helped dispel my self-doubts. After conversion was official, I received start-up funds, an invitation to New Faculty Orientation, and introductions to campus resources. I began to present my research on using creative writing pedagogy in composition courses and vice versa, at conferences. I also had more time and support to publish my fiction and nonfiction. It was no longer something I did on the side but an integral part of my professional identity. I felt more like a teacher, writer, and member of the academy than ever.

2017–2019: Pushback from above

Although we (the five of us) didn't understand the big picture at the time, the grounds for managerial resistance to conversions appear (in retrospect) to be of two sorts. First, although our university population had been growing steadily for years, it was becoming clear that we couldn't keep doing so, and that our overall system population was dropping, both of which resulted in pressure from the system to be more conservative about staffing. Second, another department on our campus also converted most of their NTT faculty into tenure lines during this period and then saw their number of majors drop badly, which left them with a problem management calls "tenure saturation": not enough courses to fill out everyone's workloads without running small (therefore expensive) sections. While that's unlikely to be a problem in an English department with two required general-education composition courses, it's no surprise that management was concerned about it.

ALISON: I moved to Pennsylvania from Washington, DC, in 2013 and spent four years struggling to find stable work. Unable to gain traction in higher education after holding a full-time teaching position at a for-profit school in Northern Virginia, I broadened my search to private high schools, alternative educational institutions, and even took the first course toward a secondary education teaching certificate. With two small children to support, I was being encouraged by family and friends to abandon teaching for something that would give me less angst and more security. So far removed from my creative writing training and practice but not quite passing as I had hoped to as a writing studies specialist, I felt stuck.

Suffice it to say, when I signed a one-year contract with WCU in August 2017, I all but collapsed with exhaustion and gratitude. I would be teaching a handful of composition courses, a business writing course, and a technical writing course. I wasn't aware of 11G when I was hired but later came to understand that some of the most supportive and welcoming faculty I initially encountered had been converted through 11G. I aspired to join them, but conversations with them about the policy always seemed kind of taboo, covert . . . definitely off the books. I was discouraged from being hopeful about my prospects. I am, however, difficult to contain when I set my sights on something, so I stopped asking about 11G and instead focused on my craft, the department, the students, and proving myself quietly. I dove deeply into the philosophical and pedagogical structure of our First Year Writing program and worked to build and shift my course approach to match. Being in a department with such a strong and inclusive writing studies faculty helped me develop my claim to writing studies and confidence to teach a sophisticated writing curriculum. I was in love with my job, and 11G gave me tenuous hope for my future.

NANCY: Like Alison, I was offered a one-year contract to teach creative writing and first-year writing. Before my interview at WCU, I was aware of 11G, and this knowledge along with the fact that WCU has a strong creative writing program affected my decision to take a job located hours away from where I still live. It was my understanding that the English Department conducted national searches in order to hire instructors with the credentials and experience to teach full-time for more than one year at WCU. I assumed I was in it for the long haul, that those of us hired in 2017 were among a small group of adjuncts who could attain 11G status, depending on our evaluations. Immediately after I was hired, I sensed there was tension about 11G conversion, but I didn't know much more.

I also perceived, for the first time in my life, that my creative writing background and degrees were important. My first year at WCU, I taught both creative writing classes and FYW classes. I felt very lucky to teach these classes as an adjunct. During that year I felt like my professional identity and personal identity (my experience teaching FYW and my creative writing background) were beginning to intertwine. Maureen and I worked closely on committees; Lisa and I bonded over poetry. When Lisa read from her collection of poems to a large audience, I experienced an "aha moment." Here was a published poet *and* a successful, tenure-track writing studies professor.

SETH: The good news in late summer 2017 was that we had converted and begun tenuring people. Faculty doing the hiring that summer were excited about great new people.

Early in the fall, I could feel something wrong. In conversations with department and union leadership, I gathered that management was backtracking on what had been clear approval to hire terminal-degree-holding faculty into full-time positions with an eye toward conversion. I didn't understand why. Eventually I had my first real conversations with Alison and Nancy about how the union might (and couldn't) help. Once I realized that management was resisting full-time hires to avoid conversions, I got mad, and nervous; grievances about adjunct hiring rarely win; their "temporary" (contract term) status makes demands for longevity and stability tenuous, no matter how rational they are. Meanwhile, I knew that our union leadership was negotiating *something* for the newly hired people, and I needed to stay quiet.

LISA: As part of the contingent faculty moving through 11G into possible tenure, I was on the precipice of a sea-change in the department in terms of the organizational structure, the emotional life of the department, and my place in the scheme of things. In theory, 11G seemed clear to everyone else I suppose, but I was on the outside of the process even as my life would be significantly impacted by the outcome. Essentially, there seemed to be a deep-seated fear among the tenured faculty that the 11G would create more problems than it solved. I am certain they wondered whether the adjuncts who had been working for years teaching the general education and first-year writing courses could also maintain a high standard of instruction and scholarship across the department and university. Would we respond to the offer of 11G as consummate professionals? I believed I could, and I knew my peers in the adjunct office who had been tested and evaluated at every turn could rise to the occasion, we just needed the chance to "claim" our place among our peers in the field. Proclaiming myself a writing professor in my department really depended upon acceptance from my peers as much as the university administration.

ALISON: In my second semester, rumblings began that 11G was going away, and that adjuncts would be kept at less than full-time to eliminate conversion opportunities. My chair called a meeting to let the adjuncts know that, starting in fall 2018, we could only be offered three classes. This was a death blow to hopes of conversion. My fragile grip on stability was upended, and I was devastated.

I went to the chair privately. I wanted him to know that taking away the opportunity for TT was like asking adjuncts *not* to care, *not* to try harder, work harder, be better. I wanted him to know that taking away the opportunity would reduce us to room fillers, hot air blowers, baseline dwellers. It's in my DNA to care deeply about my work, so I suggested making the qualifications for 11G more rigid. Anything to keep hope alive for the contingent faculty eager to excel in their craft, to focus energy in one university and one department instead of cobbling together a full load through multiple universities. Now I realize I was naïve to think my chair had any control over any of this. He was closing his first year as the trial-by-fire department chair and was simply following orders. I certainly felt cornered by this 11G policy, which was designed to help and empower someone like me but instead threatened to keep me from protected, full-time work.

That is when I decided that I needed to get a PhD. I needed more of my life to be under *my* control. I needed to make deliberate decisions that had the power to put the question of stability in *my* hands. Don't get me wrong, as a middle-aged single parent of two, also being a full-time doctoral student comes with enormous compromises, and student debt, but at least they were mine to make. I made peace with the fact that my MFA was not enough of a credential. It did not have the power I thought it did. I knew a PhD could open more opportunities, gain me more respect, and give my voice more authority, whether at WCU or elsewhere. I felt like I needed to abandon my creative writing self and fully embrace writing studies as a single identity. I enrolled in a doctoral program in literacy that was fully online and began working to make that happen.

NANCY: During my first year at WCU, I taught CRW classes and served on the CRW Committee, which helped me feel like an active member of a vibrant creative writing program. Before coming to WCU, I had not taught CRW to college students in years, and suddenly, during my first year, I was planning readings and helping my CRW students with poetry manuscripts. Maureen and the CRW faculty welcomed me. I felt respected and needed from day one.

After the meeting in spring 2017, however, I lost hope in attaining 11G conversion. Like Alison, I was devastated. We were both warned that we most likely wouldn't get a full-time schedule if we remained at WCU. As an adjunct, I knew there were no guarantees, and yet I had quit my teaching jobs near my home and moved my family from the DC area to West Chester for the academic year. I felt naïve; as contingent faculty, I should have known better. The faculty held meetings with new adjuncts in an attempt to answer questions about our future at WCU. Alison and I talked extensively about our plans during this time. I remember telling her my teaching career was probably over while she said she was getting a PhD. I suddenly felt I should have made the same decision fifteen years ago.

The First Year Writing director genuinely asked me what would make my situation as an adjunct at WCU tenable. I immediately told him I wanted a non-contingent position with benefits (of course) teaching any subject I was qualified to teach in the English Department. I explained that I didn't need to teach creative writing ever again. After all, my pedagogical approaches to composition and creative writing were intertwined, and I enjoyed teaching first-year writing. But I surprised myself when suddenly, out loud, I had said I didn't care about creative writing in the sense that I would invest all of my time and energy to FYW pedagogy in exchange for stability, even if that meant taking a multi-year contract without hope of conversion. Assuming my creative writing scholarship wouldn't count on paper, I thought I would need to start over, so to speak, as a "writing studies scholar." I would attempt to refashion my professional identity without knowing what that meant.

I continued, however, to receive full-time teaching loads, which included CRW classes. Although I didn't know if I would be hired to teach from one semester to the next, I committed to more service work and fully immersed myself in the CRW Department and community. My effort was not a means to an end during these years. My involvement sustained me and reaffirmed what I knew all along: I would never be able to separate the poet from the writing teacher. I would never be able to fully pivot. In fact, the more I invested myself in the CRW Department, the more confident I felt about my place at WCU English Department, even if I taught composition for the rest of my career and even if my career there was short-lived.

MAUREEN: Nancy says her work for CRW sustained her, but she helped sustain the CRW program, too. Though she was claiming writing studies, like many in our department, she was also maintaining a dual professional identity. Her enthusiasm and experience were incredibly valuable to our CRW students and to me on the CRW Committee. I felt uncomfortable asking for her help when I knew adjuncts were not compensated for service, but Nancy never made me ask. She always volunteered. As Lisa has mentioned, I knew adjuncts were doing the work. At the same time, I felt that I was fulfilling the requirements for TT, but that it could still be taken away at any moment. Watching what happened to Alison and Nancy reminded me of the tenuous nature of our situation. Despite these doubts, I also sensed less resistance to 11G from the department for a combination of reasons. The first tenured 11G faculty were successful. They took leadership roles in writing studies and proved to be worthy members of the department. New faculty without entrenched ideas about 11G were hired, including two for whom I even served on the hiring committee. I had significant support and expectations from the CRW Committee. Some faculty members felt the issue had been debated enough. On a personal note, between July 2016 and December 2018, I had two babies and took two semesters of leave, so I might have missed some 11G discussions. The tone in the department felt more accepting and collegial, but the shift might have been internal, too. As I accrued more years of experience and balanced this with pregnancy symptoms and new parent exhaustion, I had less energy physically and mentally for the angst of the early 11G process.

SETH: We had hired full-time people for years. We developed an ethical hiring policy that satisfied our department. We converted ten people; three had been tenured and/or promoted. Maureen's leave meant she missed conversations about concerns for MA-holding candidates in the tenure process; so far, they've all done well. From a labor perspective, the system was working the way it was supposed to, and I didn't understand why anyone would break it.

Things have recently become a little clearer. Once WCU was under the 25 percent cap, management reasserted their preference for hiring terminal-degree holders into TT positions and part-time NTT positions. They worried about staffing flexibility in light of unstable enrollments. Rumor—third-hand—has it that the *US News and World Report* college ranking system, which penalized institutions for TT faculty without terminal degrees, was an issue. None of those affects the conditions in which Alison, Nancy, and other full-time instructors were working. We won arguments for full-time schedules for some faculty, but we had to do it every semester even though needs hadn't changed, and the faculty were teaching well.

2020–2021: Nancy and Alison Were Given a Promise of Continued Full-Time Employment and Formal Recognition of Their Place in the 11G Process; Lisa and Maureen Get Tenure

The problems over the last couple of years have been mitigated—not solved, exactly, but at least remedied to some degree. Lisa and Maureen are both successfully tenured and promoted. Alison and Nancy got commitments, negotiated with our dean and provost by our department chair, director of First Year Writing, and Seth, to maintain their full-time workload through the timeline that allows them to get to the 11G conversion vote. That agreement has potential ripple effects, one of which is that we will no longer offer full-time teaching (except in emergencies) to faculty without terminal degrees—but since MFAs are terminal degrees, the new stance doesn't reduce the complexity of the situation for creative writing MFAs hired to teach primarily composition courses. But it's better, at least, for people who are here, and it clarifies (im)possibilities for new hires.

ALISON: I have been at WCU for eight consecutive semesters, and although anticipating an employment contract each semester has been stressful, I have been grateful to receive one. During the drafting process for this project, my chair told me that the administration has committed to employing me through the process to 11G. This news was exciting, and I felt relief. I felt that faith in me as an instructor was renewed, and I felt validated that the work I am doing is important to the department and administration.

Of course, I questioned for more than a moment all that I had invested, physically, monetarily, emotionally, and intellectually, in two years of a PhD program. Was it for nothing? Would everything have turned out just the same if I hadn't started the program? I had doubted it for a decade, but it turned out that my MFA *was* enough. But, when I made the decisions I made, my future at WCU was uncertain and I have to keep reminding myself of that. I feel incredibly lucky to be in the position I am and working for an institution that wants to, and tries very hard to, do the right things for the community, the environment, the students, and the faculty, whether it is expensive or difficult. I guess I can say the same for myself, and I know having a PhD will improve my work in the classroom and allow me to do important research in writing studies. Although I began claiming writing studies when writing studies began claiming me (starting in grad school and my work as a TA), having a PhD in literacy completes that transformation, which is important. I get great enjoyment from teaching FYW courses, and I think I am needed there. I know in my heart I'm a writer and I am hopeful that the stability conversion to TT may bring can provide a balance in identities similar to what Maureen and Lisa experienced.

NANCY: Like Alison, the chair's surprising news that I would receive full-year contracts for two consecutive years, enabling me to apply for 11G conversion, was surreal. In fact, that news has been pivotal in reaffirming my identity as a creative writing scholar. In the past, my research in my field has been for my students, not for a promotion. For the first time in my twenty years of teaching, I feel like my CRW background is an asset for my students and my career. The English Department, my peers, and my students have gotten to know my poet-self and have, with the prospect of 11G, given me an opportunity to have a future in a non-contingent position at WC. I used to ask myself if I was a creative writer who happened to teach FYW. Right now, I don't ask myself that question. With the help of tremendously supportive people and hope in the 11G "system," I'm working it out.

MAUREEN: When we began to write this, I was also writing my tenure and promotion application. The success of Lisa and other colleagues gave me hope. Reflecting on my decade at WCU showed me I had followed an unplanned, but not uncommon, trajectory. My ability to teach both CRW and FYW had been an asset to the department, especially during times of significant faculty changes within the CRW program. I was able to produce scholarship in FYW and devote more time to my creative work. Since my Statement of Expectations included both general education and creative writing, this production was required for tenure and promotion. The security of tenure-track status also meant I no longer needed to teach at many schools. Colleagues in both FYW and CRW welcomed my service. While being a creative writer in First Year Writing had sometimes felt like a secret I should hide, I now had years of teaching, scholarship, and service that showed I was effective in both disciplines.

In 2021, I received tenure and promotion. Friends and colleagues struggling at other schools, and even the adjuncts at WCU, remind me there is still so much more work to do to ensure that our students are taught by professors who are valued members of their departments and universities. Supporting faculty who commit to FYW, regardless of their origin stories, strengthens the discipline and helps students.

LISA: During the entire two-year process of applying for tenure, I sat across from Maureen's desk (she has been my officemate through five years of adjuncting and tenure track) and wondered whether I would still have a chair in the office the following semester. I received my official tenure letter in December 2019 and my promotion to assistant professor in June 2020. It's an understatement to say that I was in shock. Not because I hadn't done the work required to earn tenure, but like so many of my 11G colleagues, there was a sense that we had not earned the right to sit at the table without the terminal degree. One emotion that kept surfacing through the process was anger—that the labor I had invested and experience I have gained teaching since 1997 *must be worth something*. Now that I've had a few months to enjoy being awarded tenure, I am ecstatic, but not for the reasons you might assume.

I spoke recently to colleagues describing my new place in the university as moving from "looking for advocacy" to "looking to advocate." When Adrienne Rich speaks of *The Dream of a Common Language* in her book of poetry of the same name, she isn't saying we should all speak English, she's asking the reader to imagine a world where the traditional *codes of communication are broken open to allow for all of the voices that have been silenced for far too long*. In the poem "Power," Rich talks about scientist Marie Curie as having denied "her wounds came from the same source as her power" (quoted in Popova 2015).

Essentially, the honor of getting tenure has transformed my sense of purpose. I don't mean to sound like someone who has experienced an awakening altogether and sees none of the horrific reality that exists around contingency, but I am genuine when I say that I see the only way to create the kind of university we want is to push forward where we can at every opportunity. As an adjunct I spent decades trying to work in my field in a way that could sustain my need to teach and still pay my bills. As a poet and creative writer, I stood on the sidelines too, allowing the prevailing norms and expectations around writing to dictate my own process. I also lacked the time I needed to focus on my writing. Essentially, creative writing seemed adjunctive—a luxury. That too is a mindset I intend to shed. Again, I needed to see that exploitation of labor—mine and countless others—was keeping us from having any chance at reaching our potential as writers and scholars. So out of the decades-long journey to gain a place in the university, I learned that labor is the fulcrum around which everything else moves. That understanding is, for me, the most important I have acquired regarding the life of an academic.

Closing

In the fall of 2022 and with great relief, Alison and Nancy, along with a third colleague, were successfully converted to TT. Although this hard-fought battle was won, the victory offers no guarantee for future negotiations. Even with policies protecting tenable labor conditions, sometimes departmental and institutional cultures push back. In our department, even labor-conscious faculty helped create the complex conditions for Lisa, Maureen, Alison, and Nancy, who all faced some version of this problem: how (and how much) compromise is necessary to sustain identities as creative writers while claiming enough identity as writing studies specialists to meet the department's and institution's expectations. In chapter 9 of this volume, Raymond Rosas lauds the welcome that writing studies often extends to new members from other fields; our experience has been that, as creative writers, our welcome has been, well, less welcoming. Without recognizing it, the department pushed us to claim or be claimed by writing studies when our original credentials were enough to be hired. The extent to which the four creative writers felt obligated to move away from (or background) our creative writing identity differs, and the extent to which each feels authorized to or hopes to reclaim it does too. What matters is being made to think about it at all.

Ethnographic accounts make generalizations hard to draw because of the specificity of the details. That said, our experiences suggest recommendations/cautions for departments that hire creative writers as writing instructors, especially when they are asked to refract—if not change—their identities in the name of labor equity. The recommendations boil down to this: Invite creative writers to claim writing studies as a professional home but minimize the pressure to do it primarily (if not solely) as a defense against losing their jobs. Better pay and job security are obviously good, but don't pressure faculty to concede professional autonomy that builds on the identities that faculty bring with them; and support the development of new professional identifications when necessary.

The rest of this section is our effort to detail as best we can what that looks like in practical terms. Specific attempts to take up the recommendations will obviously need to respond to local policies and conditions. We also want to emphasize the need for clarity for faculty and administrators who have to follow/execute policies. Simplicity can be deceptive.

With those points in mind, this list of possibilities is long and not close to exhaustive.

CURRICULAR AND PEDAGOGICAL ISSUES
- Pedagogical autonomy in composition courses so creative writing faculty can maintain connections to their identities and teach to their strengths.

MAUREEN: By necessity, I used creative writing pedagogy in my composition courses, but when I found them successful, I researched the extensive history behind this and ended up presenting my experiences at several conferences.

ALISON: I admire Maureen's perspective and bravery. I never thought to have creative writing act as a lens through which to teach composition because I thought it would be frowned upon. I worked instead to mold my approach to how I had seen writing studies faculty teach and pedagogical research I had conducted. Although I was converted to tenure-track in the fall of 2022, I'll probably stick with this until the safety of tenure is realized.

- Involve creative writers in making decisions and developing curriculum, as long as participation is compensated or credited toward promotions or conversions.

ALISON: Because Maureen, Nancy, Lisa, and I came to writing studies as creative writers and nontraditional academics, we definitely hold unique perspectives on writing, teaching, and students. I think perspectives like these could encourage healthy balance and offer diverse points of view for department and curriculum design.

PROFESSIONAL DEVELOPMENT ISSUES
- Access to professional development and clarity about what's available.

MAUREEN: As an adjunct, I overheard someone mention we qualified for funds to attend conferences. I was never told that directly. After converting to tenure-track, I received start-up funds, a new laptop, department and university mentoring, and introductions to campus resources. It felt like winning the lottery without knowing how poor I'd been.

SETH: Yes, adjuncts have conference travel money. It's less than the TT faculty get, but it's something. We need to be louder about that.

NANCY: I didn't even know adjuncts could ask for anything.
- Support for faculty required to take up a new field.

SETH: Departments with graduate programs should offer their NTT faculty access to courses, ideally toward terminal degrees (Maisto and Kahn, 2016).

Michael Murphy (2017) has proposed certificate programs that would work similarly ("Head to Head").

LISA: I completed twelve credits in the Composition and Rhetoric graduate program at Temple University between 2016 and 2018. Having an advocate in a nearby English Department with a Comp/Rhet PhD program was akin to a miracle for those of us who needed credits[2] but couldn't find a program willing to allow us to do so without matriculating. Another colleague and I were allowed to take four courses when normally the department only allowed two unmatriculated. That single act enabled me to complete twelve credits on time and become fully eligible for promotion.

ALISON: I was lucky I could be in a position to pursue a doctoral degree. Many brilliant instructors are not. Investing in faculty to gain training and support student success seems like a no-brainer.

EVALUATION AND PROFESSIONAL ADVANCEMENT
- Credit creative writing as professional activity even for creative writers whose primary teaching responsibilities are composition.

NANCY: Most of my creative and scholarly work focuses on poetry. I'm not even sure what counts for tenure.

MAUREEN: Me neither, Nancy! After conversion to TT, I focused on presenting and publishing on FYW and pedagogy, to be safe. Writers have to write, though, so I hope the fiction and nonfiction I've published will count, too.

ALISON: This would definitely help with motivation to write! It is always the sacrificial line item on my list of to-dos. It would also help me to connect the work that I claim with the work that I am claimed by, in my mind and in my soul.
- Clarity regarding degree status, including the status of the MFA as terminal degree.

NANCY: When I received my first MFA, PhDs in creative writing didn't exist. I really didn't know my MFA was considered a terminal degree when I began teaching at WCU since a handful of universities offered PhDs by then.

ALISON: Unlike Nancy, I knew my MFA was considered a terminal degree, so I never imagined I would have trouble getting a job. As the years went by, however, it didn't seem to make much difference in the hiring process for first-year writing instructors. I had begun to think, literally until this year, that unless you had a PhD, tenure track was a pipedream.

As the chapters throughout this book demonstrate, *claiming* writing studies can be powerful. As we've shown, claiming writing studies can be especially powerful in material labor terms. We hope also to have complicated those claims by showing that when writing studies *claims* faculty, or makes faculty feel coerced into claiming it, not only can individual faculty's situations become personally and professionally difficult but the ripples can affect institutional policy and culture in unpredictable ways. Recognize those complexities; be mindful of demands; create the best conditions for faculty to succeed. That writing studies can do all of those things is the best claim for it we can make.

Notes

1. For more information on the contract issues and their ripple effects, see William B. Lalicker and Amy Lynch-Biniek's (2017) "Contingency, Solidarity, and Community Building: Principles for Converting Contingent to Tenure Track."
2. Pennsylvania law mandates a master's degree plus ten credits for an appointment to assistant professor.

References

Bishop, Wendy. 1993. "Crossing the Lines: On Creative Composition and Composing Creative Writing." *Writing on the Edge* 4, no. 2 (Spring): 117–133.

Bousquet, Marc, Bill Hendricks, and Kevin Mahoney. 2009. "Stabilizing Persons, Creating New Lines." *Academe*, November–December, 2009. https://www.aaup.org/article/stabilizing-persons-creating-new-lines.

Chang, Heewon, Faith Wambura Ngunjiri, and Kathy-Ann C. Hernandez. 2012. *Collaborative Autoethnography*. New York: Routledge.

Hesse, Douglas. 2010. "The Place Of Creative Writing in Composition Studies." *College Composition and Communication* 62 (1): 31–52. http://www.bu.edu/wpnet/files/2010/09/Hesse-Creative-Writing-Composition-CCC-2010.pdf.

LaFrance, Michelle. 2019. *Institutional Ethnography: A Theory of Practice for Writing Studies Researchers*. Logan: Utah State University Press.

Lalicker, William B., and Amy Lynch-Biniek. 2017. "Contingency, Solidarity, and Community Building: Principles for Converting Contingent to Tenure Track." In *Contingency, Exploitation, and Solidarity: Labor and Action in English Composition*, edited by

Seth Kahn, William B. Lalicker, and Amy Lynch-Biniek, 91–101. Fort Collins: WAC Clearinghouse/University Press of Colorado.

Maisto, Maria, and Seth Kahn. 2016. Review of *The Humanities, Higher Education, and Academic Freedom: Three Necessary Arguments*, by Michael Berube and Jennifer Ruth. *Academe* 102, no. 3 (May/June).

Mayers, Tim. 2005. *(Re)Writing Craft: Composition, Creative Writing, and the Future of English Studies*. Pittsburgh, PA: University of Pittsburgh Press.

Murphy, Michael. 2017. "Head to Head with EdX." In *Contingency, Exploitation, and Solidarity: Labor and Action in English Composition*, edited by Seth Kahn, William B. Lalicker, and Amy Lynch-Biniek, 71–89. Fort Collins: WAC Clearinghouse/University Press of Colorado.

Popova, Maria. 2015. "What Power Really Means: Cheryl Strayed Reads Adrienne Rich's Homage to Marie Curie." *The Marginalian*, April 24, 2015. https://www.themarginalian.org/2015/04/24/cheryl-strayed-adrienne-rich-power/.

Scott, James. 1990. *Domination and the Arts of Resistance*. New Haven, CT: Yale University Press.

6

From Pell Grants to Tenure Track

Precarity and Privilege as a Disciplinary Pathway

CYNTHIA JOHNSON

"I smell cigarette smoke. That smell! Who is that coming from?" My high school Spanish teacher asked this loudly in front of my classmates and me, and I watched in horror as she wove around the room, literally sniffing me out. I did not smoke, but I lived in a small rental house with my parents who smoked often and indoors. When she reached me, she hesitated. She clearly expected someone different, maybe someone she could more easily discipline or accuse of smoking in the parking lot. She didn't expect me—a "good" student—quiet, obedient, white. She looked away, mumbled, "That smell!" one last time, and dropped the subject.

***

I've reflected on that high school experience for many years now. When I was younger, the feelings it evoked were confusing: fear, shame, and alienation mixed with relief, comfort, and safety. Now, I recognize it as the same position that has frequently propelled me through my education, a position of tangled precarity and privilege as a white, lower-class, queer woman. On one hand, I was being sought out, identified as inappropriate or not belonging, and my very scent unpleasant. On the other hand, I was protected, my privilege as white, silent, and obedient in itself signaling that I belonged.

I share this opening story because I believe it exemplifies the same tangled positionality that has led me and many others to writing studies, a discipline that privileges the same silent obedience and whiteness that I grew up embodying, while also demanding that its participants find a home in precarity. In the following, I share a researched narrative of how I arrived in and navigated the discipline of writing studies. I focus on how, like many others, my precarious situation made me not only susceptible to but grateful for positions of heavy labor and little reward. Yet at the same time, my privileges carried me through these opportunities in ways that many are denied access. More specifically, as I became increasingly aware of both class differences and my own queerness as a teenager, I compensated through obedience, silence, and diligence in my studies, attributes that writing studies rewarded. I am also white, had family who vehemently encouraged my education, and was often able to conceal my class difference and orientation—privileges that undeniably paved my way.

While a number of forces led me to the field of writing studies, three interrelated reasons tied directly to my tangled positionality of precarity and privilege: (1) writing studies champions professionalism; (2) it often demands uncompensated, affective labor; and (3) it requires financial precarity from many of its participants. Coming from a lower-class background, I was easily lured by the promise of professionalism. To be a professional was to be upwardly mobile, to succeed according to my blue-collar upbringing. Further, I was willing to get there through labor few others wanted but that was easily handed to me. As Donna Strickland (2011) discusses, managerial or mechanical labor is "coded as white and feminine" (46). I was also comfortable in the financial precarity the discipline requires in the form of low graduate stipends and expensive conferences; financial strains were not unfamiliar to me. In other words, I was drawn to the field in part by my comfort with poverty and uncompensated labor, as well as the looming promise of a profession.

Professionalism

Through sub-specializations such as writing program administration, writing studies consists of more than just the study of writing. The discipline is undergirded by, as Strickland (2011) calls it, a *managerial unconscious*. Strickland argues for a view of the field as "a complex economic enterprise that has almost from its beginnings demanded management as a result of its ubiquity in the ever-expanding American higher education system" (7). She further argues that as the field grew into a hierarchical organization "with a class of experts structuring and overseeing the work of a group of nonexperts," it adopted the discourse of professionalism, including "calls for control and systematization of knowledge" (58). She uses Robert Hariman's description of professionalism as the dominant discourse of the middle class, allowing them to "continue their assault upon those privileges withheld by the aristocrats above them while protecting their newly won gains from the immigrant hordes pouring in below" (cited by Strickland 2011, 55). This description captures part of the tangled conditions of precarity and privilege that lured me to writing studies. While this underlying promise of professional opportunity can be particularly attractive to those seeking upward mobility from impoverished situations, it simultaneously denies access to many.

In addition to its roots in hierarchical organization and management, professionalism in its current form consists of a code of thinking and behaving that is often synonymous with heteronormative middle-class whiteness. I do not mean to critique professionalism that encourages respect for people and the collaborative negotiation of knowledge, but rather I am critiquing the superficial forms of professionalism that police participants' language (e.g., Young 2010; Tellez-Trujillo, this volume), dress (e.g., Manthey 2019; Hull, Shelton, and Mckoy 2020), and a myriad of unstated mannerisms and rules, often referred to as the *hidden curriculum*. In "Making Places as Teacher-Scholars in Composition Studies: Comparing Transition Narratives," Resa Crane Bizzaro (2002), a Meherrin and Cherokee woman and scholar, reflects on this experience, stating, "I—too—learned to abide by rules my colleagues didn't even recognize as rules in order to remain in higher education, struggling to join in the conversation among the privileged" (498). Many of the authors in this volume likewise speak to this experience, such as Karen R. Tellez-Trujillo, who states in this volume, "The most stifling feeling was the sensation that in order to become a successful English teacher I'd have to manufacture some sort of whiteness." Students and faculty from disenfranchised backgrounds often are not privy to, or they choose to reject, the unstated rules of professionalism that are built on a "white racial habitus" (Inoue 2015, 45; Byrd, this volume), a peer-enforced gatekeeping mechanism in the academy.

Uncompensated and Affective Labor

With professionalism comes different forms of labor beyond the work of the discipline. In addition to the expected and rewarded teaching, research, and service, there's the *unexpected* and *uncompensated* teaching, research, and service. There's also the affective and emotional labor that disproportionately affects historically marginalized scholars who are navigating a discipline that feels inaccessible, unwelcoming, and inequitable. In her oft-cited book *The Managed Heart: Commercialization of Human Feeling*, Arlie Hochschild (1979) describes emotional labor as the self-policing people, and especially women, do to fit into a workplace or to create a comfortable environment for colleagues and consumers. She describes emotional labor as "the management of feeling to create a publicly observable facial and bodily display" (7). In a discipline heavily defined by professionalism, emotional labor weighs most heavily on those unaccustomed to the white middle-class culture they are forced to navigate.

Multiple forms of uncompensated labor affect participants at all levels of the discipline, as indicated by the expansiveness of academic guilt ("I should be working"), burnout ("This is too much"), and imposter syndrome ("I'm not good enough") reported in the study by Dana Lynn Driscoll, S. Rebecca Leigh, and Nadia Francine Zamin (2020). They found 60 percent of surveyed faculty and 89.6 percent of doctoral students in composition studies regularly experience academic guilt; 54.3 percent and 68.5 percent, respectively, experienced burnout; and while the researchers didn't ask specifically about imposter syndrome, 40 percent of interview participants mentioned it. However, it's important to note that these widespread syndromes are not synonymous with the forms of emotional and affective labor felt by members of the discipline who lack representation, community, and systemic equity and support. Bizzaro (2002) notes, "Due to lack of minority role models, many emerging professionals—including myself—feel further isolated and marginalized by a lack of power" (495). In this volume, Khadeidra Billingsley describes in detail the "labor of tokenism" she experiences as she navigates graduate work as the only Black woman in her program, and Raymond D. Rosas recounts being asked, on sight, "are you supposed to be here?" For many of these scholars, uncompensated labor in the forms of research, teaching, and service come with the additional affective burden of tokenism and assimilation. Further, many attempt—and are often unfairly expected—to "change academia from within by serving as role models and mentors for those who follow us" (Bizzaro 2002, 497). Genevieve García de Müeller and Iris Ruiz (2017) (see also Perryman-Clark and Craig 2019) document this work at the WPA level, finding that "WPAs of color feel they are most responsible for working on initiatives that change the perception of race issues in college writing programs while also feeling silenced by writing programs if and when they try to advocate for race-based initiatives" (23). So, while the field is in many ways defined by its insistence on uncompensated labor, it would be remiss not to emphasize the added affective and emotional labor put specifically on marginalized scholars. While acknowledging my white privilege, I hope to use this narrative to speak to the affective/emotional labor experienced through my class dissonance, as well as the queer affective labor attached to performativity, (re/dis)orientation, coming out, erasure, and activism.

Financial Precarity

Lastly, I want to draw specific attention to the financial labor required of those in the discipline of writing studies. Coming from a lower-class background and lacking a financial safety net as I entered the field, I found my entrance into the discipline defined in large part by the debt I accumulated, the time spent calculating bills next to my barely livable graduate student stipend, the energy spent filling out loan and reimbursement paperwork, and the affiliated emotional labor of assimilating into a middle-class culture despite a continually perceived class difference. Jeffrey Williams (2006) describes how debt "plays off pleasure and pain," giving students hope while also "inducting students into the realm of stress, worry, and pressure over their precarious toehold" (165). However, despite entering the field from a place of poverty, my experience is far from unique. Graduate students who enter their programs from a place of financial security are leaving in extreme debt and often accepting contingent positions that offer unlivable wages and no benefits. Jim Ridolfo's (n.d.) work collecting job market data from 2012 to present shows that, at the field's peak in 2014, 325 writing studies jobs were advertised, while the estimated total in 2022 is 207—an over 36 percent drop. Academia as a whole has also long been documenting the increase in contingent positions, of which our field plays a significant part. The 2016 "CCCC Statement on Working Conditions for Non-Tenure-Track Writing Faculty" reported, "From 2005 to 2012, the number of contingent faculty members increased from 48.2 percent to 52.9 percent at doctoral-granting universities, held steady at about 61 percent at masters-granting universities, grew from 55 to 57 percent at baccalaureate colleges, and stayed constant at almost 80 percent in two-year colleges. One 2010 study, for example, found that roughly 75 percent of faculty were working off the tenure track, most part-time" (CCCC 2016). Throughout the following account I reflect repeatedly on the effects of my own economic background on my path into the discipline, but I also want to attend to the economic anxiety the field itself produces as its members put themselves through graduate school, professionalize through expensive conferences, suffer a dwindling job market, and strategize difficult paths to job security. In other words, financial precarity is demanded of many members of the discipline in exchange for their entrance.

The following, then, is a brief account of my entrance into the discipline of writing studies. My precarity and privilege as a white, lower-class, queer woman guided me to a field whose entrance was defined by the promise and peril of professionalism, the affective labor of assimilation, and the familiarity of financial precarity. My strategy into the discipline remained what it was throughout my adolescence: silent obedience as a form of hiding and a tendency to say "yes" to roles and responsibilities few wanted. This strategy into the discipline is not uncommon, as Bizzaro (2002), too, states, "Based upon the conditions of our current job market, I have learned the economic necessity of answering 'Yes!' immediately and emphatically" (504). Yet, what ultimately allowed me to claim writing studies as my discipline was finding the voices and spaces that brought me out of my silence and taught me to say "no."

Navigating the English Department

Recently, on the first day of an Introduction to Technical Writing course that I teach regularly, a student told me she was looking forward to my class. She told me that her major was technical writing and she was excited to find out what that was. Without having to, I asked why, then, she had declared technical writing as her major. She explained that she had initially been interested in other English disciplines, but she thought she'd have better job prospects in technical writing. I knew this or some variation of it would be her answer because I'd heard it before and because I'd done it myself.

The midsize state university that I attended for my undergraduate degree did not have any equivalent to a composition or rhetoric degree. They did, however, have literature, creative writing, and technical writing degrees, along with a general English degree. I remember sitting with an academic advisor during my college orientation and explaining that I was interested in an English major, but I was nervous about job security. The advisor quickly steered me toward technical writing, citing the program's job placement rating. I was in no place to turn down a promise of employment, so despite not knowing what technical writing was, I unenthusiastically agreed to what sounded like the straightest, and consequently unqueer, path toward job security. (Now I happily recognize technical writing for its many queer, subversive, and divergent paths.) This was the first of many influenced decisions that led me to writing studies. As is today true of most students, my career path has always been determined in part by ardor and in part by economic anxiety, or as Chase Bollig (2015) says, "'the good man speaking well' is looking for a job after graduation" (151).

While "Technical and Business Writing" maintains its own classification code within the National Center for Educational Statistics (NCES n.d.), this code exists within the larger grouping of "Rhetoric and Composition/Writing Studies." So, while the two fields are often treated as separate, they maintain a close relationship and many students and scholars find their way back and forth between the two. However, technical writing has a unique advantage in attracting students suffering economic anxiety. Rather than falling into the "literacy myth" that improved writing and communication skills alone will provide upward mobility (as Bollig [2015] contends of writing studies), technical writing often focuses more specifically on positioning students within the workforce in ways they can navigate. In "What's Practical about Technical Writing?" Carolyn R. Miller (1989) argues that both composition and technical writing are "practical" insofar as they serve as means to ends. She states, "Freshman composition aims to help students be more effective as students, technical writing aims to help them be more effective as engineers or accountants or systems analysts" (14). While this simplified account doesn't hold true across all contexts, many students—and university advisors, it seems—see technical writing as a field that more directly connects students' education to their entrance to and navigation of the workforce, thus alleviating some level of economic anxiety.

This experience points to a larger issue, which is that students coming from impoverished situations often feel they can't choose educational paths based on interest or inspiration, but rather must choose paths they believe will help them pay bills and achieve upward mobility. Jeffrey Williams (2006) argues that "debt teaches career choices" (164), pointing to the increasing number of business majors: "This is not because students have become more venial or no longer care about poetry or philosophy; rather, they have learned the lesson of the world in front of them and chosen according to its, and their, constraints." As a result, the emotional toll placed on both students and their instructors magnifies as they pursue education not as a means of growth or development, but as a "consumer service" (Williams 2006, 163) meant to provide them financial security. While I now have found a home in technical writing as much as writing studies, my path into the field was paved with emotional resistance and economic anxiety.

Finding Writing Studies

While technical writing was my first step toward writing studies, writing center work served as the actual gateway. During my first year of college, I survived financially through Pell Grants, living with family, and working in an O'Reilly Auto Parts warehouse, where my mom still works third shift. But when my car, three years older than me, began to stall, I looked for work at the university and discovered the writing center. Like many opportunities I pursued in academia, my privilege as white, quiet, and studious afforded me the job; however, as in many of those same opportunities, my motivation for the work was different from many of my colleagues.

I often found while working as a writing consultant that I wasn't treated as a worker struggling to pay bills, which I was; I was treated as a student or intern who should be grateful for the professional experience. I struggled to reconcile these identities, both of which were true on some level. While the minimum-wage pay was low, I *was* grateful for a part-time job on campus that acknowledged my role and schedule as a student. However, a desire for experience wasn't the reason I took on as many hours and responsibilities as possible, despite struggling to balance the work with my classes. Rather, my motivation for this work, which was perceived and rewarded by my supervisors as ambition and drive, was mostly financial insecurity and an attempt to conceal my class dissonance. Again, I learned early on to say "yes" to roles and responsibilities that few wanted, believing it kept me hidden or even proving that I belonged.

My experience working in the writing center wasn't entirely one of discomfort, however. In many ways, it was also one of the first places in academia where I felt I belonged. In my major, I often struggled with expectations of professionalism. I found that my way of thinking, speaking, and behaving always felt awkward or clunky as I learned the unspoken rules for how to navigate academic spaces. When I was once asked to dress in "business casual" for a class presentation, I spent hours researching what that meant and figuring out not only how to afford it, but how to move comfortably in it. I also was learning to edit documents for "correctness," which in my particular program once included developing a portfolio in which I corrected "errors" found in public places, regardless of context or dialectical diversity. But in the writing center, I was learning to combat those exclusionary forms of professionalism. We were encouraged to actively resist hierarchy, normativity, and presenting ourselves as professionals. Instead, we were peers working to collaborate with other students—not correct them. As Harry Denny (2010) has pointed out, "writing center practitioners must queer the dynamics that put forth particular codes of identity and intellectual practice as 'normal' and others as not" (49). So, while I struggled to make sense of my class identity in my new white-collar, part-time job, I also began easing into my queer identity and orientation toward the world around me. I found meaning in the collaborative work I was doing with other students and their writing, and when I learned the university offered an MA in rhetoric and composition, I began to see a new path for myself.

Graduate School Conditions

When I learned that the university where I was completing my bachelor's degree also offered a master's degree in rhetoric and composition, as well as an assistantship that would cover my tuition, allow me to teach, and pay me a whopping annual stipend of eight thousand dollars, I eagerly pursued the professional and financial opportunity. I encountered two groups of people in that program: some for whom eight thousand dollars was a comfortable supplemental income, and some, like me, who made it through the program with some combination of the stipend, unideal living conditions, and additional work or student loans. Once again, the promise of professionalism and upward mobility encouraged me to say "yes" to what was in front of me.

I turn again to Bizzaro's (2002) narrative, in which she describes her experience working through graduate school: "Two days each week for five years, I drove six hours round-trip to take graduate courses toward my PhD; three other days each week, I cleaned houses and worked part-time as a secretary at the university where I am now marginally employed" (487). These conditions are not uncommon for graduate students in writing studies, and programs themselves do little to acknowledge the conditions under which their graduate students are living, with teaching, research, service, and professionalization expectations rising to meet an increasingly competitive job market. As a result, Liz Miller (2020) is one of many who has documented the "detrimental effects of overwork, economic and career instability, and isolation, amongst other factors, upon [graduate student] mental health" (1). (See also Banville et al.'s [2021] discussion of graduate student precarity in technical communication.) The academic guilt, imposter syndrome, and burnout reported by Driscoll, Leigh, and Zamin (2020) begins before participants even become full members of the discipline.

Further, the financial labor that goes into navigating graduate school is an under-discussed labor. The time spent calculating expenses, the energy put into securing and often concealing employment outside the university, and the ongoing labor of navigating student loans is too often the defining pathway into the discipline. Yet in order to escape these scenarios, graduate students must find ways to excel despite these conditions. As Bizzaro "learned the economic necessity of answering 'Yes!' immediately and emphatically" (2002, 504), my colleagues and I likewise learned this strategy early in our graduate school careers. We joined professional organizations, sought freelance editing and design work, took on low-paying administrative positions, and proposed to conferences that we couldn't afford. When it then came time for my MA colleagues and I to apply to doctoral programs, many of us were devastated to learn the cost of graduate school application fees that would limit which programs we could apply to. Several of my colleagues chose, instead, to pursue adjunct or alternative work rather than go further in debt for entrance to a field that seemed so intentionally inaccessible to them. For these reasons, it's unsurprising that according to the 2013 National Census of Writing, only 7 percent of respondents at four-year institutions identified as people of color, 8 percent identified as LGBTQ, and less than 3 percent identified as disabled. The labor and privilege it takes to gain entrance into writing studies makes it a discipline that simply is not accessible to many people, particularly those from marginalized and disenfranchised communities.

Job Market Labor

The fifth and final year of my doctoral program was difficult for several reasons, the most prominent being what Caroline Dadas (2013) describes as "the process of officially *entering* the profession through the job market" (67, emphasis in original). The professional, emotional, and financial labor that went into this process was unparalleled.

Like many, by the time I reached the academic job market, I was at the peak of my burnout and debt. What's more, my graduate program only guaranteed funding for four years, with the possibility of a fifth year—despite nearly all students requiring at least five years to finish their degrees. For me, that translated to the privilege of a one-semester fellowship, followed by the precarity of unexpectedly becoming a "part-time instructor" for the final semester of my program. This included a nearly 25 percent pay cut, a requirement to contribute to the state's teacher retirement system, lost access to student loans, and the beginning of my loan repayment grace period. Once I was actually on the job market, I then had the privilege of going on several campus visits, for which I had to front the money for travel, lodging, and sometimes even meals, and then spend hours organizing receipts and completing paperwork for reimbursement that often took months to come. I again share this experience not because it's uncommon but because it's incredibly common—privileged, even. In some ways, I believe my deep-seated economic anxiety served as an advantage over some colleagues who had never experienced this financial strain before. Severe economic stress as graduate students put in the required labor for the job market is part of our pathway into the discipline. And while this pathway isn't unique to writing studies, it's a broken system that we continue to participate in and perpetuate.

The affective labor of the job market reaches beyond that of financial strain. Jennifer Sano-Franchini's (2016) research into the emotional labor of the job market demonstrates how the necessary "performances of intimacy [required of candidates] are highly contingent on culture, race, gender, class, (dis)ability, and embodiment" (11). The same precarity and privileges that had carried me through academia continued to carry me through the job market. While I obtained interviews and campus visits, the familiar feelings of class dissonance and (dis)orientation followed me as I fretted over how to dress, how to act, and even how to eat. Restaurants, privileged spaces that I often felt excluded from in my childhood and early adulthood, became critical sites of the job market. I also experienced a unique form of labor identifying as queer on the job market. I had done notable service for LGBTQ+ organizations during graduate school and proudly highlighted that work in my application materials. I was fortunate to be in a place where I felt comfortable with this identity. Even so, the process of interacting with strangers and describing the nature of my service work brought tension to my body. At one university I visited, the job committee took measures to discuss this work with me and signal the slowly growing acceptance in their campus community. While this conversation was well-meaning, it weighed on me.

In the end, I was privileged to secure a tenure-track job, but in the process, my research and teaching suffered, my budget suffered, and my mental health suffered. As entrance to the discipline, I was being forced to abandon the purported values of the field to focus on presenting myself and my work in a professionalized way. And while I also had many positive experiences on the market—including polishing my research; meeting many kind, welcoming scholars; and working closely with my mentors—I'm acutely aware that "particular bodies are privileged on the market" (Sano-Franchini 2016, 111) and mine was, mostly, one of them.

Pathways

The pathway into the discipline for many members is affective labor and financial precarity. For many others, there is no pathway at all. While these disciplinary problems are part of a larger university system that exploits university workers for surplus labor, we remain far too complicit. However, what ultimately allowed me to claim writing studies as my discipline was encountering the voices within the field who were working to change it. While a promise of professionalism first drew me in, the opposition and activism are what kept me here. As my critical consciousness grew, I renegotiated my relationship with the field, finding new spaces and pivotal strategies. Graduate students' and underrepresented faculty's pivotal strategy into the discipline shouldn't be quietly saying "yes" to everything but finding the spaces and support systems that allow us—encourage us—to ask questions, speak out, and say "no." In order to pave more inclusive and equitable pathways into the discipline, and in order for graduate students to claim writing studies as their own, we must continue to grow these resistant, subversive spaces. In this final section, then, I highlight three truly pivotal strategies that allowed me to claim writing studies as my discipline.

SEEK SCHOLARSHIP THAT FIGHTS DISCRIMINATORY PRACTICES

Throughout my entry into the discipline, I struggled continually with expectations of professionalism that were new to me and often unspoken. I found comfort, though, in the many scholars who were working to dismantle policing forms of professionalization and to redefine professionalism entirely. First, in the writing center, I found comfort in resisting hierarchy and a focus on idea-building over correctness. Peer collaboration rests on the understanding that "no student is wholly ignorant and inexperienced. Every student is already a member of several knowledge communities" (Bruffee 1984, 644). Writing studies purports to recognize and value everyone's unique expertise and meet them where they are. And while we very often lose sight of this value, its presence is a large part of why I'm here. Moving beyond the writing center and into the classroom, the antiracist scholarship of Vershawn Ashanti Young (2010) and Asao B. Inoue (2015), and so many others, gave me models for what that kind of equity could look like in my classrooms.

As I continued seeking my pathway into the discipline, I also continued seeking the work of scholars speaking against the discriminatory practices I was witnessing. While on the job market, I was attuned to Sano-Franchini's (2016) work on the emotional labor of the job market and Caroline Dadas's (2013, 2018) work on the job market's inaccessibility. More recently I have found comfort from the scholars resisting professional dress in the discipline, such as Katie Manthey's (2019) "Dress Profesh" project and the work of scholars such as Brittany Hull, Cecilia D. Shelton, and Temptaous Mckoy (2020), who, in "Dressed But Not Tryin' to Impress: Black Women Deconstructing 'Professional' Dress" argue, "Because minority bodies are always, already under scrutiny and subject to explanation and qualification, they are often conditioned to be aware of and responsive to the presumed standards of professionalism just to survive." They then challenge the expectations of professionalism in both language and bodies in the academy. These conversations, even when not speaking directly to me or my experience, painted the picture of a discipline I wanted to belong in.

POSITION YOURSELF IN SUPPORTIVE SPACES

As can be seen in the section above, much of the work being done to increase equity and inclusion in the discipline is by people of color, women, graduate students, junior scholars, and others who are disenfranchised or underrepresented in academia. It goes without saying that affective labor is rampant in the field, disproportionately affecting historically marginalized people and those without institutional protection. As a lower-class queer woman, I often struggled with many of the normative middle-class conventions of the academy, but what provided me relief and support were the spaces—formal and informal—that create a dialogue about the hidden curriculum and critique the unequal affective labor of the field. Digital spaces like Twitter, job market Facebook groups, Reddit, and listservs like NextGEN serve as spaces where graduate students can observe conversations about the hidden curriculum; ask informal questions (such as how to dress for the job market); and critique the discriminatory practices they see. Organization member groups and SIGs (special interest groups) also act as this resource for many graduate students and junior faculty. Additionally, scholarship such as Miller's (2020) work on forging "networks of care" in graduate school and Pamela VanHaitsma and Steph Ceraso's (2017) work on establishing feminist horizontal mentoring as early career academics offer further footholds into the discipline for new and emerging members. For me, participating in my university's Graduate Student Pride Association became an important step in discovering the spaces and work I wanted to be a part of in the field. While finding these supportive spaces is in itself a form of labor, it's a pivotal strategy that can help protect those who can't find that support and protection from the larger discipline.

BUILD NEW PATHWAYS

While the previous two strategies serve almost as coping mechanisms for those entering the discipline, this third strategy works toward change. The discomfort of my financial precarity continues today, even as I now hold a tenure-track position. The debt collected through entrance into the field follows many of its members well into their careers, if not indefinitely. Further, those coming from lower-class backgrounds, like myself, continue to struggle with the professionalized middle-class culture of academia, even when we ourselves become a part of that middle class. While many of the economic issues of writing studies—such as underpaying graduate students and overreliance on contingent faculty—don't originate in the field but in the neoliberal university, we remain far too complicit in these practices. As Sano-Franchini (2016) says of the job market specifically, "Even though global politics and economies certainly limit possibilities, we are not excused from being complicit in partaking in what are arguably oppressive hiring practices" (119). Part of why I claim writing studies as my own, then, is the desire to pave new pathways for those who follow.

The work of change has historically fallen on those already disenfranchised by the university, and so any calls for change must start there. Established and privileged members of the discipline must do the work of paving more equitable pathways for those who follow. I see this work being done in ways that would have helped me by scholars such as William P. Banks, Matthew B. Cox, and Caroline Dadas (2019), who use their scholarship to queer pathways into writing studies and professional writing. I also recognize the work of scholars who address the financial precarity and debt of their students. Doing so can help normalize the place of the lower-class in academia and more specifically help these students find a home in writing studies. James Rushing Daniel (2018) argues that "the writing classroom, as a part of the university, is itself implicated in the funding apparatus of higher education that will leave many not only in debt but psychologically defeated as well" (199). In response, he suggests compositionists "guide students through critiques of neoliberal conditions, debt specifically, in the hopes that students will eventually strive for financial justice in public contexts" (203). For many like myself, acknowledgment of my class difference came only when I was in a position to demand it, and never when I needed it most.

While I hope to further contribute to these paths connected to my own identity and experience, privilege must be leveraged to support those whose paths, or lack thereof, look different from our own. Antiracist practices that acknowledge diverse language, dress, and ways of knowing and being must extend beyond our first-year writing classrooms and into our graduate programs and hiring practices. Our conceptions of professionalism must be made transparent and based outside a white middle-class culture, and hiring committees must base their decisions in considerations untainted by ableist practices (in regard to both the hiring process itself and research/publishing expectations for graduate students and contingent faculty). Further, we must continue to make spaces for emerging members of our discipline to safely ask questions, speak out, and say "no" with the appropriate protections in place. These spaces must be made not only in our institutions, organizations, and social media but also in our journals and citation practices. And while none of these suggestions are new or mine alone, our progress in this work indicates they bear repeating.

Just as in my opening story, I'm still always waiting to be sniffed out as unbelonging in the discipline. And as a white, formerly lower-class, queer woman, I'm aware of my proclivity toward silence and subservience as a way of hiding and belonging in academia. I believe these attributes are what paved my path into writing studies, where my quiet labor was often rewarded. I accepted the uncompensated labor that was offered to me, and the financial hardships that many in the discipline face felt familiar to me. I realize, too, that this is a privileged path into the discipline; that silence and hiding is a privilege; and that the struggles I faced are exacerbated for so many others. What made me truly claim writing studies as a discipline, however, was that it also brought me out of my silence. It offered me subversive voices and spaces that allowed me to say "no" to uncomfortable forms of professionalism, to uncompensated labor, and to traditions that perpetuate these exclusionary practices for others. Writing studies self-reflects; it makes space for scholars, like those I've cited here, to call for change; and on its good days, it listens.

References

Banks, William P., Matthew B. Cox, and Caroline Dadas. 2019. *Re/Orienting Writing Studies: Queer Methods, Queer Projects.* Logan: Utah State University Press.

Banville, Morgan, Meghalee Das, Katlynne Davis, Allison Durazzi, Evelyn Dsouza, Emily Gresbrink, Elena Kalodner-Martin, and Danielle Mollie Stambler. 2021. "Identity, Agency, and Precarity: Considerations of Graduate Students in Technical Communication." *Programmatic Perspectives* 12 (2): 5–15.

Bizzaro, Resa Crane. 2002. "Making Places as Teacher-Scholars in Composition Studies: Comparing Transition Narratives." *College Composition and Communication* 53 (3): 487–506.

Bollig, Chase. 2015. "'Is College Worth It?' Arguing for Composition's Value with the Citizen-Worker." *College Composition and Communication* 67 (2): 150–172.

Bruffee, Kenneth A. 1984. "Collaborative Learning and the 'Conversation of Mankind.'" *College English* 46 (7): 635–652.

Conference on College Composition and Communication (CCCC). 2016. "CCCC Statement on Working Conditions for Non-Tenure-Track Writing Faculty." National Council of Teachers of English. https://cccc.ncte.org/cccc/resources/positions/working-conditions-ntt.

Dadas, Caroline. 2013. "Reaching the Profession: The Locations of the Rhetoric and Composition Job Market." *College Composition and Communication* 65 (1): 67–89.

Dadas, Caroline. 2018. "Interview Practices as Accessibility: The Academic Job Market." *Composition Forum* 39.

Daniel, James Rushing. 2018. "'A Debt Is Just the Perversion of a Promise': Composition and the Student Loan." *College Composition and Communication* 70 (2): 195–221.

de Müeller, Genevieve García, and Iris Ruiz. 2017. "Race, Silence, and Writing Program Administration: A Qualitative Study of US College Writing Programs." *WPA: Writing Program Administration—Journal of the Council of Writing Program Administrators* 40 (2): 19–39.

Denny, Harry. 2010. "Queering the Writing Center." *The Writing Center Journal* 30 (1): 95–124.

Driscoll, Dana Lynn, S. Rebecca Leigh, and Nadia Francine Zamin. 2020. "Self-Care as Professionalization: A Case for Ethical Doctoral Education in Composition Studies." *College Composition and Communication* 71 (3): 453–480.

Hochschild, Arlie. 1979. *The Managed Heart: Commercialization of Human Feeling.* Berkeley: University of California Press.

Hull, Brittany, Cecilia D. Shelton, and Temptaous Mckoy. 2020. "Dressed But Not Tryin' to Impress: Black Women Deconstructing 'Professional' Dress." *The Journal of Multimodal Rhetorics* 3 (2). http://journalofmultimodalrhetorics.com/3-2-hull-shelton-mckoy.

Inoue, Asao B. 2015. *Antiracist Writing Assessment Ecologies: Teaching and Assessing Writing for a Socially Just Future.* Fort Collins, CO: WAC Clearinghouse/Parlor Press.

Manthey, Katie. 2019. "Dress Profesh: Deconstructing Power through the Clothing." *Hyperrhiz: New Media Cultures*, no. 21. https://doi.org/10.20415/hyp/021.m03.

Miller, Carolyn R. 1989. "What's Practical about Technical Writing?" In *Technical Writing: Theory and Practice*, edited by Bertie E. Fearing and Keats Sparrow, 14–24. New York: Modern Language Association of America.

Miller, Liz. 2020. "Mental Health in a Disabling Landscape: Forging Networks of Care in Graduate School." *Xchanges: An Interdisciplinary Journal of Technical Communication, Rhetoric, and Writing Across the Curriculum* 15 (1). http://www.xchanges.org/welcome-to-issue-15-1.

"National Census of Writing." 2013. Writing Census. Accessed January 31, 2020. http://writingcensus.swarthmore.edu.

National Center for Educational Statistics. n.d. NCES. Accessed January 31, 2020. http://nces.ed.gov/ipeds/cipcode/cipdetail.aspx?y=55&cipid=88369.

Perryman-Clark, Staci M., and Collin Lamont Craig. 2019. *Black Perspectives in Writing Program Administration: From the Margins to the Center*. Urbana, IL: National Council of Teachers of English.

Ridolfo, Jim. n.d. "Rhet Map: Market Comparison." Rhet Map. Accessed June 20, 2023. http://rhetmap.org/market-comparison.

Sano-Franchini, Jennifer. 2016. "'It's Like Writing Yourself into a Codependent Relationship with Someone Who Doesn't Even Want You!' Emotional Labor, Intimacy, and the Academic Job Market in Rhetoric and Composition." *College Composition and Communication* 68 (1): 98–124.

Strickland, Donna. 2011. *The Managerial Unconscious in the History of Composition Studies*. Carbondale: Southern Illinois University Press.

VanHaitsma, Pamela, and Steph Ceraso. 2017. "'Making It' in the Academy through Horizontal Mentoring." *Peitho Journal* 19 (2): 210–233.

Williams, Jeffrey. 2006. "The Pedagogy of Debt." *College Literature* 33 (4): 155–169.

Young, Vershawn Ashanti. 2010. "Should Writers Use They Own English?" *Iowa Journal of Cultural Studies* 12 (1): 110–118.

7
Finding Resilience in Writing Studies on the United States–Mexico Border

KAREN R. TELLEZ-TRUJILLO

Part I

I went to school with this girl who started every frank conversation with the question, "¿Sabes que?" which translates to, "You know what?" and then she'd let you have it. Whether it was a judgment or a demand for an apology, she got right to the point. Over time, I got used to her approach and even grew to respect it. There was no beating around the bush with her, and her friends always knew where she was coming from. I've tried being more direct in conversation but have not really mastered the abrupt greeting that my friend had made her own. In the spirit of getting right to the point, and after a lot of thought, I decided that I want to start this chapter by telling you about the big event that almost changed the course of my path to a PhD in rhetoric. There are a couple other important stories, but I want to jump right in, so, here it goes.

¿Sabes que? I almost left the master's program at my Southwest Border university because one of my professors decided to call me and my writing out as ugly in a graduate-level modern rhetorical theory class. This professor had done other things to insult me and many of my peers, but he really did it this time. Most of his insults centered on language and identity, and while

I thought his comment would send me running from the master's program, it eventually sent me straight into the writing classroom. What he did to me strengthened my commitment to completing an MA in rhetoric and later the doctoral program in rhetoric. His comment also opened the door to research of enactments of feminist rhetorical resilience (Flynn, Sotirin, and Brady 2012) by my students. But before we celebrate the good stuff, this is how it all went down.

It was the third time I'd taken a class with this well-known, white, male professor in my department. The class was deep in a conversation about the preferred writing styles of the many white men in an early edition of Patricia Bizzell and Bruce Herzberg's *The Rhetorical Tradition*. As a way of making an example of the opposite of what these early men of rhetoric valued in writing, he pointed out in front of my peers in a graduate rhetoric seminar that I wrote/write "like a Mexican." My writing was his example of the disorganized kind, lacking clarity and being nonlinear.[1] This comment shocked and embarrassed me. I was so stunned that I asked myself, "Did I just hear what he said?" The most honorable end to this story would be to say that this event was met with a warrior-like response from me, but Malinalxochitl, goddess of the desert snakes and scorpions was nowhere to be found. I didn't fight back. All I could muster was, "I wish you'd be more of a teacher and less of a scholar. We already know you know this stuff." Burn. Not. After this random comment came an ashamed silence followed by the sickening sensation that I wasn't safe in this class regardless of how many of my supportive peers were present. I was in my sixth year as a student in the English Department and as Cynthia Johnson notes in this volume, I had already learned that there was compensation for "obedience, silence, and diligence in my studies" in this department, and as Johnson also mentions, these are "attributes that writing studies rewarded" (chapter 6).

For these and many other reasons, there was no striking or stinging on this day or any day for a long time after my professor called me out. I sat face forward in my chair, my feet hot, hands tingling, and head hollowed—stunned, insulted, and full of questions. I imagined this was what hypertension feels like. I knew this professor had been annoyed with me and a handful of other Latinx women in class a week prior when we argued in response to his notification that we were naturalized to like Gloria Anzaldúa. He consistently talked negatively about Anzaldúa and reminded us that he didn't care for her work. The clincher was that in his opinion we were naturalized to like her only because we were Mexican women. As if annoyance wasn't enough, he'd moved to insults.

This professor used me as an example of bad writing, while also denigrating my people, as well as the other two women in the room who identify as Mexican American and bilingual. Instead of fighting back, I retreated into a self-effacing space filled with questions like, "What does a Mexican write like?" "What does that even mean?" and "Does writing like a Mexican mean I can't succeed in writing studies, or college?" As I mentioned before, I had been in two of this professor's classes prior, once as an undergraduate in a technical communication course and in another graduate course. I wondered if he had always held negative opinions about my writing and if I had ever been safe in his classroom. These are the kinds of things one wonders when one is accustomed to accommodating others' discomfort, practiced in passing off negative comments as playfulness, or as innocent chiding. As time moved on, the focus of my hurt and anger became less about my feelings and more about my wonder as to how many more people have been pushed to the margins with comments such as this one and have given up, accepting that they cannot write themselves white. The welt placed on my tan face by this professor lasted, remaining red, tender, and swollen for years. It would flare up every time I picked up my dissertation for revision or prepared a conference paper. I have always struggled to write "fluently" and know that everything I write requires numerous revisions, but this feeling was something different. Every academic paper started to look like a collection of poorly patched together quotes and paraphrases, moves I used to avoid using my own words, or express my views, until enough comfort was established to allow my opinions to surface again. I felt like a "pretender on the throne" with every paper I submitted and with every thought of moving forward in the MA program.[2] The most stifling feeling was the sensation that in order to become a successful English teacher I'd have to manufacture some sort of whiteness. Would I have to abandon my writing style, voice, or my relationship to the Spanish language and Border identity? Is that even possible? Why was I asking myself such ridiculous questions?

It ended up taking numerous writing assignments, help from professors, and many moments of reflection to emerge from the violence of this professor's shove outside the boundaries of white, academic discourse. This was the most remarkable event of marginalization on my path to doctoral studies but was not the first time I'd felt that I was far outside of what I thought an English major should look like or be.

The undergraduate creative writing degree at my university required that students attend monthly fiction and poetry readings. These Friday night events showcased graduate student theses in poetry or prose, as well as a guest reader who was already published and well known. The dimly lit, cavernous space where the writers performed each Friday night was filled with student readers and faculty who were predominantly white, as were the majority of the guests. This demographic of attendees is remarkable and not easy to accomplish in a city populated by 60+ percent people who identify as Hispanic or Latinx. Poems read on stage by non-Hispanic or -Latinx creative writing students recalled vacations to Korea, Budapest, and Costa Rica. Some faculty members read pages from chapters that channeled experiences in Telluride, Colorado, or recalled Christmas stories where stacks of twenty-dollar bills were set aside to stuff last-minute stockings. These early impressions of the English Department came from well written, imagery-filled evenings that did all but help me to feel as if a middle-class, thirty-something woman of color would fit in, let alone succeed in a master's or doctoral program.

I consistently managed feelings that my place was anywhere but within the field I'd chosen, which led to a lack of confidence in teaching writing to my students. This professor who had insulted me tried many times to categorize me in the classroom by asking questions about my grasp of the Spanish language, whether or not my parents are Spanish speakers, and where my people are from. I believed all along that he knew these answers as I openly identify as Mexican American, I have a Spanish surname and married name, and, to be blunt, I look like the average Chicanx woman who lives in the Southwest Borderland. Assuming that I identify as Mexican would not be a stretch or an insult. I don't believe that my identification as a Mexican American was his issue with me, but it was the only qualifier he needed to feel confident enough to remind me publicly that I am not white and that my writing doesn't do whatever it is that long-since-passed rhetoricians asked of their students. Because I don't and likely never will write in a way that meets ghostly expectations, I was diagnosed and sent to the composition sanitarium for Mexicans who don't write in a straight line. The terminal fate of suffering from a shade of brown that my professor couldn't quite name and thought could be treated with a nearly forced and impossible convalescence was possibly to stop my writing altogether. It is not as if I could sever or abandon my history, culture, or relationship to language, or writing. This amputation would certainly not assist in enriching or growing an identity that is foreign to me and, further, was not desired or even necessary. I had spent so much time trying to meet the academic standards of my teachers from elementary school through college, working to write in a white voice and style that it is no wonder that my writing is a maelstrom of words. As a pedagogue, he should have known this.

As angry as I get now upon recollection of my early experiences in my English Department, I want to make it known that my predominant emotions at that time was that I was crushed, embarrassed, and frustrated that I wasn't angry enough. My initial thoughts were that the comment my professor made was only about being Mexican, but upon reconsideration and with distance, the insult has since become more complicated. Along with being Mexican, there are the intersections of being a woman, a student, and an emerging Chicanx scholar who was finding her voice through writing. The attempts made to marginalize me were centered on my body as a whole, in an attempt to shut my mouth and stop my hand. If I could be made ashamed of my writing, maybe I would stop the action (Calafell 2010; Chávez 2018)? My eventual wish for myself was that I would resist being resilient. This feeling ended up being a gift, as my eventual anger caused a shift in focus from myself to my students. I realized how convenient it is for a professor such as mine, a tenured white man, to have resilient students. As long as this professor has resilient students, he does not have to change his behavior. For too long, professors have relied on students to fall back on the behaviors they have learned when facing adversity in writing classes that come in the way of silence, withdrawal, or adjustment to discomfort at the expense of their physical and mental health.

After a few years had passed, I began feeling hopeful that the encounter with my professor would stop showing up as nausea and would eventually transform into writing. The more I read and taught, the more I realized that experiences of marginalization like mine were present in the work of others, and most important at the time, my composition students. Learning this through my dissertation study was a reassurance that I needed to be in the writing classroom, where I could make time for writing that gave my students an opportunity to reflect and write about the moments when they, like me, might need a way to turn feeling-filled instances of adversity into words on a page. Comfort came in Sara Ahmed's (2017) words on hope, that it is "behind us when we have to work for something to be possible." Like Ray Rosas writes, there is still a benefit to "buying into the capacity" for hope, even if others in writing studies have made moves that attempt to make or help us feel that we don't belong in the discipline (chapter 9). The work that needs to be done and was done in reflection is what I needed to see this hope through. This hope is at the heart of my commitment to myself and to my students (Ahmed 2017,

2). With my eventual realization in response to data analysis, that adversity comes in all different sizes and is met with resilience, comes my introduction and story shared as I locate myself in this chapter. It is also the way I chose to begin my dissertation (Tellez-Trujillo 2020), recognizing as Gesa Kirsch and Joy S. Ritchie suggest in "Beyond the Personal: Theorizing a Politics of Location in Composition Research" that any location is "fluid, multiple, and illusive" (1995, 8) and as a way of owning the moment that almost pushed me out but instead held me in writing studies. Together, my experiences, filled with my varying subjectivities and locations, have led me to claim writing studies, as a feminist writer and professor. As I moved from undergraduate studies to a master's degree in rhetoric and professional communication, and on to a PhD in rhetoric, I figured out after too long that writing has always been my resilient response to adversity, particularly when my consistent instinct is to retreat into silence and to accept that I don't have to feel ashamed at that.

Part II

While asking questions as to the origins of feminism in general and as a presence in one's life, Sara Ahmed wrote, "A story always starts before it can be told" (Ahmed 2017). Among other lines in Ahmed's 2017 book *Living a Feminist Life*, this statement led me to consider my origins as a feminist and, more recently, the way that adversity has mapped out my path to becoming a feminist writing studies scholar. My journey has been fraught with many jolts and twists. Along the way, I was introduced to feminist rhetorical resilience (Flynn, Sotirin, and Brady 2012) and this concept opened up a framework with which I could begin to make sense of the adversities I faced as a nontraditional Chicanx woman. The work of Elizabeth Flynn, Patricia Sotirin, and Ann Brady struck a chord with me. These feminist theorists developed their "theoretical, rhetorical, and contextual" concept of resilience in response to narrow approaches applied in other fields (2012). For these scholars, a focus on "social justice, equity, care, and gender" give attention to "processes and context rather than individual qualities and behavior" (2012, 7). I was drawn to this approach to resilience, as it removes the responsibility of the individual to save themselves, and allows us to take care of each other. Feminist resilience is ongoing, and "gives voice to and supports vulnerabilities" (12). It also takes place while working toward "empowerment, growth, health, and transformation" for the sake of the individual and the communities to which they contribute (22). Also appealing to me is that feminist resilience calls for gathering resources and relationships that will present additional means to reshape possibilities, and one's self, by seized opportunities to "change shape to meet the exigencies of . . . circumstances" (9). The end is not the goal, but continuous forward movement is what matters, calling for progression through consistent, small gestures. This is far more than bouncing back, or returning to a prior state, as popular definitions of resilience teach us.

Learning more about feminist rhetorical resilience set me up to stop and think about the ways that, through writing, one can take responsibility for their reactions, particularly in situations where they may otherwise shut down, accommodate, or accept a situation as commonplace. So, now feminist rhetorical resilience is at the center of my writing and research, as are examinations of occurrences of feminist resilience enacted by students as they recall experiences of adversity inside and outside the classroom. Teaching composition has filled me with energy to help students write about the things they have wanted to say but have not been able to, as it has for so many others. Writing in this way is as much about the way students have used words to affect others as it is about the ways words have been used on them while thinking more extensively about the effects of these words. This type of writing encourages bravery and recognition of resilience enacted in response to adversity. All too often the adversity we face is in relation to language, reading, and writing when we should be using writing to confront adversity instead. One of the most valuable aspects of feminist resilience is that it is ongoing, and carries us through daunting events and relationships, long past their occurrence, and into the complicated web of adversities faced, day in and day out.

I am content where I am now with my research interests and assistant professorship at a Hispanic-Serving Institution in Southern California.[3] Sometimes, when I look back, I am awestruck by how far I have come from being a teenage mom who took a fifteen-year break between high school and completion of a bachelor's degree. I can easily pick out all of the stops on my path that either moved me closer to the place I am now or threatened to throw me off track. Although my collision with that professor in the master's program was a major upset, ¿Sabes que? My path to writing studies goes back way farther than the *fregazo* from which I still smart and has a lot to do with the fact that I have never truly felt safe in writing classrooms. Although writing studies isn't necessarily a safe place either, many people are working hard on and off campuses to complete research in their communities and dedicating long hours to make a difference for students who are writing for their lives, or, at the very least, for their identity.

I've spent most of my life negotiating what it means to be *nepantla* (Anzaldúa 1987), always in between or in the middle of something. Some of my family members swear we are only mixtures of Spanish and some other European blood such as Italian or German, others say we are only Mexican, a combination of Spanish and Indigenous blood (Moya 2003; Anzaldúa 1987, 25). A couple hundred dollars has exposed the truth presented through vague percentages marked on a colorful 23andMe pie graph revealing that we are in fact a combination of European and Indigenous blood. You know, Mexican. This identity is best described using the word "mestiza," or what Gabriela Raquel Rios refers to as "the trope of mestizaje . . . used by elites as a way to erase or marginalize not only the Indigenous but also the Black population in Latin America and the United States" (Ríos 2016). People in my world often privilege light eyes and light skin, grasping for the European bloodlines from hundreds of years past. Even one of the most beloved of my family's patron saints is blue-eyed and light-skinned. My youngest, blue-eyed brother was nicknamed Santo Niño, after El Santo Niño de Atocha, the Christ Child, by my grandparents because of his "guero" traits. I wasn't as lucky. My identity has been more of a movement between Malinche, or the daughter of Malinche.[4] I am torn between two languages, *chingada*, screwed for having fair skin and dark eyes, and for speaking and writing predominantly in English, appearing to have smeared over my color with a thick layer of caliche.[5] Somewhere, somehow, I must have culturally betrayed someone to have advanced through a Eurocentric education. ¿Que no? As if my outward appearance didn't reveal enough of a genetic struggle between Indigenous and European blood, I use Spanglish as my default language in conversations with those who are predominantly Spanish speakers. Gasp. Most Spanish speakers are kind, when they realize that this is my solution to a struggle, rather than a language skill. I wouldn't call my use of language code-switching, "the mixing of English and Spanish," as José Cano does in his essay, "Code-Switching," because that would signal attrition of a language never fully possessed. I imagine it can be entertaining to watch me shape-shifting like Métis, to conform to the cultural expectations of my surroundings, always careful to not shape-shift into a form that will lead to being swallowed or chomped by big white professor teeth (Cano 2016).

There is a fracture in my Spanish, and I spend far too much time in fear of being called "pocha," or "cultural traitor" as is defined by Cruz Medina (2016, 93–107). Sticks and stones, right?[6] Who is it that I am afraid of? But I still worry that through this fracture in my language, where wrong words emerge, and conjugation errors spill, I will bring embarrassment to myself and my ancestors because I can't fully articulate my questions, hopes, and intentions in a language that is spoken all around me, my entire life. It would be a cop out to say that that language is complicated and leave it at that. I'd like to think that the cracks in my identity have more to do with my growth in the Borderland that spans across the space I've most consistently traveled from El Paso, Texas, to my lifelong home in Las Cruces, New Mexico. This is the Borderland that Gloria Anzaldúa (1987) refers to as a "dividing line, a narrow strip along a steep edge" (3). El Paso is a city caressed by a river, separating two countries, the United States and Mexico, adjoined by Ciudad Juarez. This Borderland of my birth is also Anzaldúa's Borderland, the "vague and undetermined place created by the emotional residue of an unnatural boundary . . . in a constant state of transition" (25). Those living on the US side of the boundary, like myself, live with and among the language and cultures of Mexico from the US, where a blend of language and cultures moves in a fluid motion, back and forth, in and out of everyday lives. This mixture is common to the area, but not necessarily natural. In a less peaceful description of the Borderland, Anzaldúa refers to this space as "una herida abierta where the Third World grates against the first and bleeds. And before a scab forms it hemorrhages again, the lifeblood of two worlds merging to form a third country—a border culture" (3). This is the wound of my emersion, a geographical abrasion, the other, from my mother, who has an ambivalent relationship with speaking, as she does with silence. Once again, I stick a pin in the spot that represents a place on my path that serves as an attraction to writing studies. Going back to my mother's traumas that have come from speaking and choosing not to is a draw I feel to a field that offers a place for those who need and want to be heard but do not necessarily want to speak. That too seems to me like living in a borderland, wanting to speak, but also wanting to be quiet. Wanting to write, but not wanting anyone to read what is written. Once again, I am not pretending that writing studies is safe, but it does offer opportunities for researchers who are at their best when in search and rescue of writing that spans uncharted contexts and developing genres. Writing studies holds a place for a Chicanx feminist to continue the work she believes she's done since birth to push back against boundaries and tell an important story, even if it takes years to do so. It's a place where a professor can ask a student to think about and talk about those moments where they felt that they were walking along the *herida abierta*, and how this had made them cringe when they are asked to write about it, and why. Surely, the hesitation to revisit the discomfort, or to open wounds, goes back generations, as others have and continue to study generational trauma.

It has been a great source of pride to think that Chicana feminism has been part of me since birth. While I was growing in my mother's body, I think I practiced speaking back to all that my mother and the women before me had endured. My preparation for being a Chicanx woman in writing studies can be traced to a world of struggle and never-ending work toward social justice started by my mother, her mother, and her mother before her. Feminism is a word that spoke me into existence (Ahmed 2017). My mother's favorite story about me takes place in the first moments after my birth. It was as if I came into the world knowing that my discomfort as a girl on this earth would be lifelong. After nine months of feeling my mother's struggle, swimming in her silence, and resilience, I made my feelings known. The moment the doctor pulled me from my mother's body, up into the air, holding my feet between his fingers, with my eyes closed, back arched, and hands in fists, my skin turned angry shades of purple and red. I then screamed the kind of scream that comes from somewhere deep down inside, lasting and strong. My mom says that this sound didn't seem to be meant for the doctor, nor for her, but was the kind of yell that leads into a deep, shrilled cry, aimed at everything and nothing at all. After years of hearing stories of my mother's experiences, I now believe the scream began as the frustration that had grown from her constant work to restrain her voice for fear of violence and passed to me as an unmistakable voice of rage. "She's a girl, but she ain't no lady!" the doctor said to my mother as I wailed. From my first breath, I disidentified with expectations of what a girl's behavior should be.[7] This would not be the first or the last time my behavior would fall outside of societal expectations of a female's behavior and although I've oftentimes silenced myself, I would learn that writing was a way to let out whatever it is that is simmering within when crying out wasn't an option.

Part III

I hold the skills of Métis within, a goddess who, because of her experiences, is aware of her resources and consistently employs cunning, shapeshifting, and insight, calculating when and how to move into spaces closed to the marginalized Other. In battles against power, I evaluate my skills and resources and contribute them to the groups to which I belong, like Hephaestus, "the God of smiths and in a sense of fire" (Dowden 2007, 46). My own body, like that of Hephaestus, presents limitations after spending most of my adult life managing the symptoms of a chronic autoimmune disease. Each day brings a different struggle with pain, nausea, and waning strength resulting from chronic inflammation over which I have little control.

This part of my story is important to tell because it is an example of silence that comes about because no one is listening. This sounds like a philosophical question that asks, "If someone is speaking and no one listens, are they really just silent?" Before my autoimmune diagnosis, management of the symptoms that could only be felt and not seen masked as other ailments, so they went undiagnosed because telling doctors that I was in pain was not enough. In many ways, suffering came double as a woman and as someone who is chronically ill, trying to identify a diagnosis, in that "the category of woman has been closely aligned with the category of disability as a term that has marked deficiency and disqualification" (Dolmage and Lewiecki-Wilson 2010, 23). Many male physicians refused me as my own provider of knowledge about my illness, and as a woman with an illness that, although undiagnosed, was seen as not "normal" and thus outside the "normative matrix" that "comprises a narrow range of rhetorical ability, which is impossible to maintain; it also overlooks the ways that rhetors make use of disability as rhetorical power" (27), of which, as a woman, was nonexistent for me.

In writing studies, I found a place where conceptions of normality and ability are complicated, discussed, and unpacked, allowing for exploration with positioning and positionalities as well as exploration of how relationships are formed with those positioned differently from normative ideals. Although I still didn't think writing studies is a magical place where all can be fixed, I see the possibilities that come from writing, discussion, and reflection that I don't see happening elsewhere. Writing studies makes possible the examination of emotional and embodied responses and reactions to being marked as different, where students can think about the way that they have also marked others (Dolmage and Lewiecki-Wilson 2010; Kerschbaum 2014). Acts of taxonomizing and categorizing assist the person or institution in smothering a voice and body into submission by labeling it as monstrous, damaging and negating its credibility, and squelching its power through acts of violence (Hawhee 2004).[8]

Some of the most important contributors to the writing studies are those like me, who have stories to share about the ways that words are used by and against them. Thus, it is the misfortune, that exclusion from fields such as writing studies starts early for many who are told or made to believe that reading and writing are not for them, or not their strength. Literacy myths, and uninformed teachers help students create narratives they learn to tell about themselves as they progress through their education as to why they avoid English classes or don't like them. Language, geographic location, and ethnicity are only some of the factors that serve as barriers between those interested in writing studies and the field itself.

Many Latinx scholars have acknowledged that rhetoric and composition studies or writing studies as a whole needs decolonizing. An important example is Iris Ruiz and Raúl Sánchez's (2016) edited collection in *Decolonizing Rhetoric and Composition Studies: New Latinx Keywords for Theory and Pedagogy* that does some of the much-needed heavy lifting in the way of discussing misused words that are used against minority students. Again, in 2016, Iris Ruiz drew attention to the need to recognize the roles of ethnic minorities in *Reclaiming Composition for Chicano/as and Other Ethnic Minorities: A Critical History and Pedagogy*. These two books were published during the time I spent in preparation for comprehensive exams, and for that I'm grateful. Within the questions presented in my examination about rhetoric, literacy, and composition was the glaring truth that experiences like mine have been written about and explored in writing studies extensively. It is obvious that there are not enough faces in writing studies that look like mine when flipping through conference programs and in literature reviews, but that does not scare me away anymore. I have found courage in the knowledge that I am among the resilient who have seen adversity and continue to withdraw and reengage when I feel ready. While feminist resilient action has informed my techniques of survival, it has also motivated me to help my students spot the resilience in their own lives, particularly when it comes to their relationships to reading, writing, and language. This is where I look forward to making a difference and accept that I made the right choice when choosing writing studies as my present and future. I anticipate that there will be struggles as I grow into my role as an assistant professor dedicated to my students and to writing studies, but I am no longer pushed to the margins so easily and am not likely to stay silent for long.

Notes

1. As with any story where there are troublesome moments, I want to add the disclaimer that by and large, the faculty at the university from which I have grown are some of the most caring, hard-working, student-centered teachers this country has to offer. Alienation from a program and a field of study can come from many places, and at times in my postsecondary education, I was the leader of my journey to the margins due to insecurities

and a stern focus on the ways I did not fit in rather than seeing the ways that I did. What you read reflects the exception and not the rule. This is not to undermine the gravity of my experience but to recognize my advisor, and the majority of my undergraduate and graduate teachers in the Department of English and Interdisciplinary Studies who worked hard in dedication to my education and successes.
2. Referring to Yolanda Flores Niemann's description of the Impostor Syndrome in "Psychology" where women of color suffer "feeling increased pressure to outperform others, outthink, and outshine their colleagues, feelings of isolation, pressure to assimilate, lack of mentors, difficulty communicating with majority group members, gender discrimination, being left out of the 'old boy' network, role complexity and doubts from their majority group counterparts regarding their competency" (108).
3. At the time this is written, I am only a month into my first assistant professorship in the Department of English and Modern Languages at California State Polytechnic University in Pomona, California.
4. From Octavio Paz's "The Sons of Malinche." *Chingar*, then, is to do violence to another. The verb is masculine, active, cruel: it stings, wounds, gashes, stains. And it provokes a bitter, resentful satisfaction. The person who suffers this action is passive, inert, and open, in contrast to the active, aggressive, and closed person who inflicts it. (21).
5. A white sedimentary rock that is turned white by calcium carbonate, which binds other minerals.
6. Medina draws on the work of Villareal, Rodriguez, and Anzaldúa.
7. Per Cheryl Glenn in *Rhetorical Feminism and This Thing Called Hope*, drawing on Gloria Anzaldúa—"A term coined by José Esteban Muñoz to describe an intentional subversion of dominant expectations for being in the world" (2017, 52).
8. Debra Hawhee. Bodily Arts: Rhetoric and Athletics in Ancient Greece. University of Texas Press, 2004.

References

Ahmed, Sara. 2017. *Living a Feminist Life*. Durham, NC: Duke University Press.
Anzaldúa, Gloria. 1987. *Borderlands: La Frontera*. Vol. 3. San Francisco: Aunt Lute.
Calafell, Bernadette Marie. 2010. "Rhetorics of Possibility: Challenging the Textual Bias of Rhetoric through the Theory of the Flesh." In *Rhetorica in Motion: Feminist Rhetorical Methods and Methodologies*, edited by Eileen Schell and K. J. Rawson, 104–117. Pittsburgh, PA: University of Pittsburgh Press.
Cano, José. 2016. "Code-Switching." In *Decolonizing Rhetoric and Composition Studies*, 63–75. New York: Palgrave Macmillan.
Chávez, Karma R. 2018. "The Body: An Abstract and Actual Rhetorical Concept." *Rhetoric Society Quarterly* 48 (3): 242–250.
Dolmage, Jay, and Cynthia Lewiecki-Wilson. 2010. "Refiguring Rhetorica: Linking Feminist Rhetoric and Disability Studies." In *Rhetorica in Motion: Feminist Rhetorical Methods and Methodologies*, edited by Eileen E. Schell and K. J. Rawson. Pittsburgh, PA: University of Pittsburgh Press.
Dowden, Ken. 2007. "Olympian Gods, Olympian Pantheon." In *A Companion to Greek Religion, edited by Daniel Ogden*, 41–55. Malden, MA: Blackwell.
Flynn, Elizabeth A., Patricia Sotirin, and Ann Brady, eds. 2012. *Feminist Rhetorical Resilience*. Logan: Utah State University Press.

Hawhee, Debra. 2004. *Bodily Arts: Rhetoric and Athletics in Ancient Greece*. Austin: University of Texas Press.

Kerschbaum, Stephanie L. 2014. *Toward a New Rhetoric of Difference*. Urbana, IL: Conference on College Composition and Communication, National Council of Teachers of English.

Kirsch, Gesa E., and Joy S. Ritchie.1995. "Beyond the Personal: Theorizing a Politics of Location in Composition Research." *College Composition and Communication* 46 (1): 7–29.

Medina, Cruz. 2016. "Poch@." In *Decolonizing Rhetoric and Composition Studies: New Latinx Keywords for Theory and Pedagogy*, edited by Iris D. Ruiz and Raúl Sánchez, 93–107. New York: Palgrave Macmillan.

Moya, Paul. 2003. "With Us or Without Us: The Development of a Latino Public Sphere." *Nepantla: Views from South* 4 (2): 245–252.

Ríos, Gabriela Raquel. 2016. "Mestizaje." In *Decolonizing Rhetoric and Composition Studies: New Latinx Keywords for Theory and Pedagogy*, edited by Iris D. Ruiz and Raúl Sánchez, 109–124. New York: Springer.

Ruiz, Iris D. 2016. *Reclaiming Composition for Chicano/as and Other Ethnic Minorities*. New York: Palgrave Macmillan.

Ruiz, Iris D., and Raúl Sánchez, eds. 2016. *Decolonizing Rhetoric and Composition Studies: New Latinx Keywords for Theory and Pedagogy*. New York: Springer.

Tellez-Trujillo, Karen R. 2020. "Enactments of Feminist Resilience in the Composition Classroom: ReScripting Post-Adversity Encounters Through Writing." PhD diss., New Mexico State University.

SECTION III

Allegiance and Identification

INTERLUDE THREE

Allegiance and Identification

The essays in this section display ragged wounds. While all of the essays in this book foreground affect, amid little cuts from relocating oneself and one's discipline to the frank wounds of imposter syndrome, Billingsley, Rosas, Zepeda, and Byrd push deeply into trauma (though they don't call their experiences traumatic) and pain to locate the allegiances sustaining them. For these contributors, identification with writing studies comes with costs and with penalties exacted by societal structures and norms.

For each author in this section, language powers allegiance and identification. For Zepeda, language provokes the questions that drive her research. For Billingsley, language delimits but also reminds her of her allegiance to future, yet-to-be teacher-scholars. Likewise for Rosas, language has both maimed those he loves and sustained him on a difficult journey. Finally, Byrd recounts his (re)discovery of Black history and subsequent deepening of his own allegiances and identification.

In this section, through reflection and research, these teacher-scholars find their way to stories frank with loss but also with the long planing and shaving of their selves, some self-inflicted (Rosas and Zepeda), some societally inflicted (Rosas, Billingsley, Zepeda, and Byrd). Systems outside but also within academia complicate the possibility of opportunity, resilience, and allegiance. Every story begins in language. But whose?

8
Being the Only One

The Embodiment and Labor of Tokenism

KHADEIDRA BILLINGSLEY

I am a token. I have always been a token. I will forever be a token. And tokenism hurts.

Introduction

In her article "The Social Ecology of Tokenism in Higher Education," Yolanda Flores Niemann defines tokens as "rare persons of their demographic groups within [a particular] context, especially in contrast with majority, numerical dominants." (2020, 325). Rosabeth Moss Kanter (1977) quantifies tokenism by stating that it generally occurs when the numerical minority comprises less than 15 percent of the collective (6). However, it is far more complex than merely a numbers game. Niemann describes tokenism as "a disrupted identity, a psychological state imposed on" (326). She contends that a tokenized status is usually accompanied by three main external perceptive phenomena including hypervisibility, polarization, and assimilation. Due to a smaller numerical presence, there is heightened awareness of the token body. Thus, this leads to an exaggeration of difference to that of the numerical majority, ofttimes the white, cisgender, heterosexual male. According to Niemann, this state of being also amplifies "social identity contingencies," which are possible judgements that are based on one's social identity in a particular context or environment. Subsequently, a change in one's identity, here of the token, begins to occur in an effort to accommodate or align with the dominant perceptions and expectations of the environment in which one is tokenized.

Tokenism is commonly subjected onto Black academics by our white colleagues and peers. It is rampantly commonplace in academic spaces. Writing studies faculty and graduate students of color have shared their experiences as tokens within the discipline. Vershawn Ashanti Young acknowledges how "whites also impose performance expectations upon us" (2004, 4). Additionally, S. Tay Glover describes how being a "token diversity figure would mean an uneven [and unexpected] load of labor and trauma" (2019, 162). In my experiences as an African American female PhD student at a predominantly white institution (PWI), I can affirm such sentiments. Tokenism has afforded me many opportunities yet has also burdened me with an equal amount of physical, emotional, and professional labor. As I have matriculated through my doctoral program as a token, I have observed and encountered various layers of burden that the identity as such projects onto the minds, bodies, and souls of Black female graduate students (BFGS) like myself. Over time, I have adopted various rhetorical tactics and strategies that have allowed me to renegotiate my identity as a token and lay claim to a space and discipline that was not inherently composed for me. These strategies, explicated below, are characterized by pain and the "heavy labor" (Glover 2019, 169) that my fellow volume contributor, Cynthia Johnson, attributes to our precarious identities as graduate students in chapter 6 of this volume. Yet, instead of letting the weight of tokenism crush my strength, spirit, and success, I use it to aid and abet my concretizing of my position and footing as an emergent writing studies scholar. Although there is still much work to be done within the discipline as it relates to securing and sustaining the well-being of marginalized bodies, sharing my experiences as a token has not only been a cathartic tactic for my own personal and professional soul-searching but also a way to challenge the discipline to do better and keep trying. The anecdotes included in this chapter seek to answer and prompt readers' contemplation of the following questions pertaining to the plight of token Black female graduate students, like myself, in writing studies programs: How does one endure? What does it mean to embody such a subjectivity? What does this identity as a token who is pursuing an advanced degree look, and, more importantly, feel like?

As Niemann affirms, tokenism is not confined by space; it is an embodied experience that permeates various contexts, institutions, and systems of oppression through the vessel of the token. Henceforth, what is the labor associated with such an experience? In this essay, I work to show how the pressures of tradition for students of color to always be ready to show up and show out is a form of abuse exerted on the bodies of minority tokens. We, whose reality this is, know there are many layers and levels of labor to our identity and struggle as tokens. I argue that the definition of token BFGS labor would not be comprehensive without the consideration of our constant exertion of personal, emotional, intellectual, and professional efforts. In this chapter, I outline the various layers of labor that are subjected on me and my Black sistas who are also "PhD-ing" it out with me. I also provide strategies for how I have managed to mitigate the burdens of tokenism, oftentimes in solitude, while continuing to matriculate. For some, including myself, tokenism provides a false sense of security; a brief sense of finally making it. As time passes, this feeling grows into a sense of being needed for more than the sake of quantity; disposability seems like something reserved for "them over there." However, in the current social climate of our nation, as well as the abundance of higher education institutions where token students of color are being exploited, these feelings can quickly dissipate and be replaced by feelings of insecurity, unworthiness, and invalidation. It is important that we examine all of the aspects, factors, and influences on how one embodies tokenism and negotiates one's existence as such as we move through spaces of academia. Tokenism is a laborious experience that oftentimes requires sacrifice of one's entire existence. In providing a more holistic conceptualization of token BFGS labor, I hope to fill in some of the awareness and knowledge gaps that seem to perpetuate the ongoing tension, conflict, and misunderstanding of what it means to be "the only one."

Playing the Numbers Game

As we take a more focalized look at the context of higher education, tokenism is not a new phenomenon and is commonly subjected upon individuals of color. The expectations to pursue higher education for students of color continue to exceed the actual number of ethnic minorities that are admitted into doctoral programs in the ivory tower each year. According to a 2019 report by the Council of Graduate Schools, approximately 25 percent of all doctoral applicants that year were individuals from minority ethnic groups. Of this, roughly only 11.8 percent identified as Black. Once we break this percentage down based on gender, prioritizing female, the number falls below double digits. Although these numbers have increased from prior years, many of these students are still specks of color in a sea of white at their universities and in these spaces, and simply put, mediocrity is not an option.

Feeling as if I am always "on," I am rarely afforded the privilege of turning off or the luxury to just have a bad day that many of my cohort members possess. There have been several class days on which I have been exhausted from juggling the multiple other life obligations and responsibilities that I attend to outside of my identity as a student. Yet, during these times, I still feel as if I must perform to the highest standard despite my overwhelming exhaustion because the reactions to mediocrity are more pronounced for me compared to my peers. I can count on one hand the few days that I actively chose to not lead but follow during class discussions or to not be the first to volunteer for a class activity. Simultaneously, I can use the same fingers to count the number of times that I was questioned about the motives behind these acts of agency. I do not dispute that genuine concern propelled my professors to inquire about these moments in which I was "not being myself"; however, these interactions insinuated that my state of normalcy is excellence, which is exhausting to maintain. Truth be told, I would like to have days where I can just fade into the background among my peers and chill for a day. However, these desired days off may not be so desirable since only 5 percent of US college faculty identify as African Americans; it is apparent that there are more melanin-ated bodies than seats at the table for us. Additionally, the hypervisibility of tokenism that Kanter acknowledges does not allow me to do that as I stand out, in terms of both physicality and ability, among my classmates. This grim realization does not warrant empathy; it further intensifies the possible negative implications and consequences of mediocrity for students of color.

When you are the only one, your existence and movements become heightened. Thus, we must work ten times harder to earn our keep. Having this realization was necessary, at an earlier stage in my doctoral matriculation, to navigate and establish my disciplinary identity. And let me tell you, existing in a space that was not created or structured for your existence is exhausting and overwhelming. Yes, graduate school is a tough feat for anyone. When pursuing an advanced degree, there are obstacles and barriers that we must all learn to overcome. However, for some of us, these hardships are exacerbated by the hyperbolic pressures of existing in a tokenized state of being.

As a Black, female rhet/comp PhD student, I constantly and consistently feel the pains, on so many levels, of this excess weight. Even though I have been a token for basically the entirety of my educational journey, it was not until I began my doctoral studies that I realized how many extra pounds of expectations, tribulations, and frustrations I was carrying as a result of such identity. Currently, I am the only Black female student in my program and only one of two African Americans; these are not atypical demographics for writing studies programs. As Kanter asserts, "The word token reflects one's distinctiveness in the context and status as a symbol of one's kind" (1977, 11). Thus, being the only one is oftentimes synonymous with the ideal of "the chosen one." We, as tokens, are commonly hailed as the representatives of academic excellence for and of our racial communities and there are certain criteria by which we must operate as such. This identification as "the chosen one" typically occurs at an early stage in one's academic matriculation, which holds true for me; however, I did not become aware of its magnitude until I started my graduate studies. From that point onward, the self-exploitation and mutilation of my mind, body, and soul amplified.

My Body Is My Sanctuary, Not Your Playground

Not only does tokenism harm the institution or context in which it is perpetuated, but also the body of the token it inhabits. Academia is not conducive to the existence of marginalized bodies and students of color. Karma Chávez (2018) has called upon rhetoric and composition studies and scholars to acknowledge the work of the body in both physical and abstract rhetorical forms. As an enacted rhetorical practice of the field at large, the body has often found itself replicating practices of white, cis, male traditions. As she affirms, "The white male body haunts rhetorical practice and criticism. But only due to its presumed absence do the actual bodies of different others become significant to rhetorical invention and study" (244). Additionally, Chávez asserts that the body claims power and asks us "who and which bodies matter, become material (important and present), in this field?" (245). Certain values are attributed to specific bodies, which may or may not be valued in a particular context. Therefore, the body is not only a site of power but also a breeding ground for judgement, particularly when it is not the norm or standard. Trying to appease the expectations and desires of others is a laborious task. Young describes several burdens of racial performance, including the requirement to "quell and fulfill the racial, class, and gender fantasies that others have . . . and that I have of myself." (2004, 16). This repeated exploitation of the tokenized body, by others as well as ourselves, necessitates an overwhelming amount of energy to labor through. When I say that I am tired, I am tired. At the onset of my program, in my first meeting with my advisor, I shared my determination to finish my degree in three years. I was not shocked by her response as she tried to iterate that there was no reason to be in a rush and that I may be taking on more than I could handle with such an overambitious goal. Yet, her concern fell on deaf ears—remember, excellence and overachievement are ingrained in the sinews of my bones. I should have listened. Taking extra courses to finish earlier than my peers, signing up for leadership roles to stay present and relevant in the minds of department administrators, while also trying to maintain somewhat of a personal life has definitely taken its toll on my body in more than one way. My weight seems to be on a neverending rollercoaster, I have lost count of how many times I have visited my neurologist with fear and certainty that I must have a brain aneurysm judging by the regular daily headaches. It has gotten to the point where I almost break in rejoicing dance if I wake up without pain. I am highly convinced that the severe headaches that I have been enduring since the beginning of graduate school can be attributed to the immense pressures that I carry as a result of my constant tokenized state. Yet, for me, these pains seem to be quite literally a necessity, a survival tactic for me to solidify my place in this space.

Make That Money, Don't Let It Make You

Anyone pursuing a graduate degree faces the threat of precarity; however, for token BFGSs, the danger is constantly looming. Financial labor can be characterized as the efforts that Black female graduate students are forced to make when attempting to stay above the poverty line, which many are not able to do. The financial burdens placed on these individuals as a result of their educational matriculation are multifaceted. In a study of nineteen BFGSs, Johnson found that, in addition to the typical graduate student financial conundrums, her participants also struggled with financial constraints and inadequate financial support systems (Johnson 2001 ctd. in Robinson 2013, 158). Some of us are mothers, sometimes in single-parent households, or we are responsible for caring for other family members. Due to the incongruence between the labor that they are exerting and the pay that they are receiving, many BFGSs are financially strapped, thus requiring them to exert more effort, oftentimes outside of the classroom, to make ends meet. Thus, a strategy that I, and other BFGSs, have utilized is to stretch ourselves across multiple responsibilities to seek out multiple means of income. More and more are having to acquire second and third jobs—or four, like me—to supplement their wages. During my first year in my PhD program, while fulfilling my GTA (graduate teaching assistant) requirements of teaching two classes, I also adjunct at a nearby university, taught English online to overseas ESL learners, and served as a substitute teacher for the local public school system. My managing of these similar yet different professional roles for the sake of survival exemplifies the "balancing act" that Raymond D. Rosas elevates in his chapter in this volume as a by-product of disciplinary identification for many graduate students. Rogers, Eaton, and Voos assert that this choice, or more like a necessity, could possibly interfere with students' ability to focus and/or complete their coursework and degrees (2014, 507). Although, according to a Bureau of Labor Statistics 2016 report, the average stipend was $30,900 nationally (BLS 2017, 193), most students do not receive even close to that, as wages greatly vary depending on one's home institution and field of study. Despite it being highly frowned upon and even more controversial, race and gender are sometimes the deciding factors in determining the amounts and allocations of student stipends. In their piece "Mapping the Margins and Searching for Higher Ground," Green et al. tell the story of Hope, a first-year PhD student who was admitted to a doctoral program at a PWI. During her first week of classes, she discovered that, although she and many of her peers of color received a tuition waiver and a partial minority scholarship, not one of them received full funding; this seemed to be reserved only for her white colleagues. Now Hope could have definitely filed a complaint against her department for discriminatory practices and had sufficient evidence to do so; however, she did not. After a conversation with her mother who advised her to basically "put her head down and continue to persevere" (Green et al. 2018, 310), she accepted the status quo.

This decision on Hope's behalf could have been largely motivated by the words of her mother, but it also seems to be rooted in a greater issue of the lack of self-advocacy and monetary negotiations for token Black female academics. Studies have shown that females, particularly those of color, are less likely to self-advocate, which is one of the most necessary skills to successfully and optimally navigate not only graduate school but the entire landscape of academia (Vo 2012, 94). As is evident in the dialogue between Hope and her mother, oftentimes we are simply grateful for having a position in the classroom, whether as a student or educator, thus we do not feel entitled to more, a trait that does not escape our male or white counterparts. This hesitance can have long-term repercussions in the form of lost employment, overlooked fellowship opportunities, and delayed degree progression and possibly completion.

A disturbing consequence of this unfortunate reality for Hope and many other BFGS tokens is that we commonly have to try to work through these issues alone, either by force or by choice. For the sake of "saving face" I kept many of my financial stresses to myself. In fear of not affirming cultural stereotypes and prejudices, I did not want anyone to know of my struggles. Lord knows that we Black women do not like to ask for help, and I am no exception. However, I knew that something had to change because the weight of my financial burdens, and those four jobs, were becoming too much to handle. I began to seek out financial resources, not only in my department but also in my discipline in the form of scholarships, grants, and awards; you name it, I applied for it. I also started to have candid conversations with faculty members whom I felt safe with about my current predicament in hopes of finding some resolution together. I did not want to further perpetuate the invisibility of tokens that Leonie Smith describes in "To Be Black, Female, and Literate," so I began to speak out and seek help. The financial woes of graduate students, which Williams calls the "dirty little secret of higher education" (Williams qtd. in Patton 2012), would benefit from exposure as it is important to consider how such a phenomenon could complicate one's educational matriculation, particularly those of minority female tokens. With more and more students having to worry about how they will financially survive another month or where their next meal will come from, it adds another layer of stress, anxiety, and exhaustion that does not couple well with the pursuit of an advanced graduate degree. This layer of labor, which is oftentimes not known or visible to school administrators, seems to warrant higher priority in the minds of all.

I'm Not Gonna Cry. At Least Not Too Much

As Carmen Kynard (2016) affirms, higher education can be a violent space, particularly for Black and Brown bodies. As tokens, existing in the spaces of academia is sure to be emotionally overwhelming. This emotional labor occurs when Black female graduate students are suffering from internal strife or mental feelings or conditions that could potentially cause disruption in their normal activities, including the pursuit of their degrees. This layer of labor for graduate students may not be visible to those around them, however it does have a profound effect on their ability to function in their everyday tasks. For the most part, I felt fine during the first two years of my program. I boasted, at times, to my friends and family about the incongruence between my experiences and the horror stories that I heard prior to starting. However, once I started working on my dissertation, it was like the floodgates opened and never closed again. S. Tay Glover speaks of how, during several moments in her graduate career, she "blinked back tears of sadness and anger" (2019, 157) as she navigated the entry spaces of academia. For me, it was the exit that hit me like a ton of bricks. Being one of the only Black, female, and lesbian individuals in her graduate cohort caused Glover to experience feelings of imposter syndrome, fear of being outed as not being worthy, and anger to snowball over time. Pressures from faculty members, advisors, family, and students, coupled with the normal requisites of life, seemed to amass very quickly and lead to feelings of wanting to call it quits. Glover's testimony starkly resonated with me. As I neared the finish line, there was a two-week period where I cried every single day. It got so bad that I visited the Counseling Center on campus for an emergency session because I literally felt like I was on the verge of losing my mind. Feeling as if I was not yet the expert that I was expected to be after writing a dissertation, along with the lack of employment that all my peers and professors expected me to quickly secure, became overwhelming and no longer suppressible. This emotional rollercoaster ride did not prevent me from finishing and asserting my identity as an authority in my field, but it did make crossing the finish line more of a drag instead of a sprint.

For many ethnic minorities, particularly African Americans, mental health is taboo and is not publicly or openly discussed among peers, especially those who are not family members. Thus, it was extremely difficult for me to reveal the emotional turmoil that I was enduring with those closest to me. However, I knew that if I did not share, I would not survive. Thus, I made the active choice to let go of some pain, which actually ended up uncovering more hurt. In most cases, family members and peers who are not in the world of academia do not understand the demands and requirements of graduate school, therefore they may not understand that it is possible for one to develop serious mental health problems during the journey. This led me to develop a sense of resentment toward my family for quite some time. The stress of having to worry about whether one will be able to pay their bills at the end of the month, if they will be able to successfully complete their coursework, the demands of publishing, as well as teaching demands can escalate very quickly and snowball into a major issue. The polarization that may isolate tokens from their colleagues may transcend to their relationships with their family and friends, which leads to self-nurturing of emotional and mental strife. Deciding if one is comfortable enough with their family, friends, or peers to admit that they may be dealing with a mental health problem can take a lot of courage and labor, which is piled on top of the financial and intellectual pressures that they are also dealing with as a graduate student. Thus, I prefer to deal with my emotional strife on my own. My strategic attempt to be vulnerable and honest about my experiences with those closest to me was accompanied by more pain that I am just not ready to work through at this point in life. Thus, in order to cope with the traumas of graduate education, I engage in extensive self-reflection as a survival tactic I work to *rewrite the trauma narrative* (Godbee 2020, 37), a technique borrowed from Black queer feminist sociologist Eric Anthony Grollman.

This process requires that I list "every challenging, offensive, and potentially traumatizing event or condition" that I can recall and then rewriting them to include moments of "pushing back," "defining [your] career for [your]self," or "defying mainstream expectations." Although for some, it may be reminiscent of reliving one's trauma, I find this experience to be extremely cathartic. Not only does it allow me to purge through the process of writing down the traumatic incidents, experiences, and/or feelings but it also allows me to have it for reference if I desire to confront individuals who may have perpetuated or caused such trauma or share my pain with those I feel safe with. I look at it as my form of self-care (in addition to binge-watching real crime television). As we continue to ponder the labors of tokenism, it is important to remember that it is not solely confined to the classroom or even the institution; it is a disease that can deteriorate one's mind, body, and soul as well.

Intellectual labor includes academic pressures and strains that a BFGS places on herself to maintain an identity that seems to be in line with the ideal student scholar. For many Black women, "education, we are told, is the key that opens the imaginary door to success" (Smith 2005, 183). This assumption is our guiding light through our educational careers. Graduate school is not an easy feat, which may explain why only 1.5 percent of the US population has a doctoral degree. With most PhD programs taking between five and seven years to complete, there is an immense amount of effort that is required on an intellectual level. PhD students are usually identified as being high-achieving students who excelled academically in their undergraduate careers (Wollast et al. 2018, 143). Thus, it would seem plausible that these students are able to meet the standard of cognitive engagement that is expected of traditional graduate students. However, for many students, particularly minority tokens, this is not the case. As a Black, queer, and female graduate student, Glover textually illustrates the "difficulties [of] successfully completing tough coursework" (2019, 162). The caliber of work, reading requirements, discussion-based and -led classes, in addition to publication pressures, caused her to constantly doubt her abilities and place in these academic spaces. Culturally, because of the high esteem associated with higher education, BFGSs place an immense amount of pressure on themselves to ensure that we are not failing our families and our home communities. In her article, "Spoketokenism: Black Women Talking Back about Graduate School Experiences," Robinson describes this perpetual feeling of not wanting to be the "Black person who came in here and failed" (2013, 170), experienced by many of her BFGS participants as many of them felt that they oftentimes carried "the weight of the Black race" (2013, 170). As the only one in many of the educational spaces that we inhabit, BFGSs commonly assume, or are given, the identity of the representative of an entire ethnic community and, as such, we feel the need to represent. However, this desire, or necessity in some cases, brings along an undue amount of mental and intellectual exhaustion.

Even on top of all of the aforementioned, BFGSs still have to endure other intellectual demands as tokens. In their piece "The Perpetual Outsider: Voices of Black Women Pursuing Doctoral Degrees at Predominantly White Institutions," Marjorie C. Shavers and James L. Moore III describe a study they conducted to gauge the effects of graduate school on the well-being and academic persistence of BFGSs. Tokenism was one of the main strands in their characterization of "the perpetual outsider," a feeling with which most of their participants identified. As they described their experiences, many of the women discussed the discomfort they felt in spaces where they had to "speak up and become the designated representative for Black women." (2019, 220). These types of occurrences not only strip these individuals of any choice they may have to assume such a role as the spokesperson for an entire diverse and non-monolithic group of people but they also require a level of extreme pressure and anxiety as they have now been deemed arbiters of the Black experience. In these moments, I choose to speak up because I would rather be the spokesperson and provide a somewhat accurate representation then for someone else, who is not qualified by either melanin or mentality, to completely misrepresent or exploit my experiences. Yet, the burden of the extensive intellectual maneuvering that is required to negotiate the tensions of one's heightened visibility and invisibility as an individual in these moments is heavy but still frequently carried by token BFGSs.

For those students who may not be or feel up to par with their peers and colleagues in their program, this may lead to sentiments of not belonging or inadequacy. Oftentimes, this may lead to students suffering from imposter syndrome—the belief or feeling that they do not belong in academia or are not meeting the standard that has been set for their being there. Glover speaks on experiencing constant "policing of [her] affects and critiques by professors and peers [and] immense stress due to conflicts from structural and interpersonal gendered racism and homophobia" (2019, 162). She describes how her only saving grace from falling victim to imposter syndrome and her ultimately deciding to leave graduate school was an African American mentor with whom she could identify and to whom she could express her thoughts and ideas without judgment. Many students are not as lucky as Glover was and are left to navigate their milieu thoughts and the politics of academia. This particularly holds true for international, ethnic minorities, and female graduate students. Deciding to pursue an advanced degree is an endeavor that requires a certain level of mental, physical, and intellectual willpower. The constant demands of one to perform in the classroom and develop innovative research and pedagogy that will benefit generations to come adds another layer of labor that is not light in weight. Thus, when developing a comprehensive definition of token BFGS labor, it is one that demands high priority and consideration.

Concluding Remarks

Being a Black female graduate student is not an easy feat by any means. Even those us who successfully complete the journey—by the time this piece is published, I will be Dr. K—endure layers of wear and tear on our minds, bodies, and souls just to get to the finish line. The strategies and tactics that I have utilized have been fraught with pain but seem to be necessary given the goal at hand. I hope that the preceding explanation of the different layers of labor that affect the ways in which we exist and navigate these spaces, that were not created with our thriving in mind, will greatly inform future scholarship and interactions. There is a dearth of research that speaks to the experiences of token BFGSs in our field, particularly that which prioritizes who we are in an effort to force universities and colleges to accommodate our existence, rather than us being accommodating in the ways that we exist. Yet, I have hope. I hope that as more BFGSs illuminate their experiences, our discipline will listen, first and foremost, and then begin to make the necessary actions to do better.

Fostering a comprehensive definition and conceptualization of BFGS labor would be beneficial on several different levels. Literature that collectively addresses all of the facets of this concept in one text seems to be nonexistent at this point, which adds to the necessity of this particular work. Several pieces address problems faced and labor endured by various subgroups of graduate students; however, as we galvanize and move toward negotiation of rights and benefits, I feel that we need to do so as a united front, which will require a comprehensive outline of our struggle as a whole. Developing this comprehensive definition of our labor would allow us to do just that. It would also have several implications for both graduate students and higher education administrators. Therefore, acknowledging the work of graduate students, even prior to officially being recognized as a member of the academy, is integral in acquiring a full picture of what it means to be a BFGS and the multiple layers of exerted labor. If administrators were able to better understand the layers of labor endured by Black female graduate students as a result of their educational matriculation, then they could possibly better comprehend the necessity for better benefits and learning environments. What's more, they would have a better idea of what it means to be a Black female graduate student and the many layers of labor that such an identity encompasses.

Tokenism hurts. Those who make it to the end are usually the ones who learned how to navigate the complexity and chaos that characterizes academic spaces (Green et al. 2018, 298). I am a token and that will most likely never change. However, I determine how I choose to exist as such. I get to choose the amount of wear and tear I put on my mind, body, and soul as well as that which I allow others to subject upon me. Coming to this realization did not happen overnight and it was preceded by a shitload of pain. It is important for us to find ways to balance rhetorical embodiment of subjectivities such as tokenism. Furthermore, we need to share them with our sisters (and brothers) who are also on the same journey; "after all, sharing is healing" (Young 2004, 2). bell hooks affirms Black women's obligation to take more initiative to share their academic experiences and survival strategies in both written and oral media (1989, 61). Thus, I acknowledge my vulnerability as a token student but also my simultaneous obligation as a future token faculty member to support those who pursue the same journey to PhD land. This piece is a response to Staci Perryman-Clark's "call for a more critical look at Black women's intellectual processes and experiences within the academy" (2009), specifically in writing studies programs. I hope that this is just the beginning. It is important for us to shine the spotlight on ourselves and make our institutions, departments, and colleagues aware of the unhealthy ways that tokenism can manifest, ruminate, and cause harm beyond repair. I hope that telling my story and strategies for enduring will help other token BFGSs who have similar—or dissimilar, because we are not all the same—experiences to begin to share. Because, at the end of the day, there is no one better to tell our stories than us.

References

Bureau of Labor Statistics (BLS). 2017. "Occupational Employment and Wages, May 2016: 25–1191 Graduate Teaching Assistants." Occupational Employment Statistics. US Department of Labor. https://www.bls.gov/oes/2016/may/education_2016.xls.

Chávez, Karma R. 2018. "The Body: An Abstract and Actual Rhetorical Concept." *Rhetoric Society Quarterly* 48 (3): 242–250. https://doi.org/10.1080/02773945.2018.1454182.

Glover, S. Tay. 2019. "'Black Lesbians—Who Will Fight for Our Lives but Us?': Navigating Power, Belonging, Labor, Resistance, and Graduate Student Survival in the Ivory Tower." *Feminist Teacher* 27 (2): 157–175.

Godbee, Beth. 2020. "The Trauma of Graduate Education: Graduate Writers Countering Epistemic Injustice and Reclaiming Epistemic Rights." In *Learning from the Lived Experiences of Graduate Student Writers*, edited by Shannon Madden, Michele Eodice, Kirsten T. Edwards, and Alexandria Lockett. Logan: Utah State University Press.

Green, Dari, Tifanie Pulley, Melinda Jackson, Lori Latrice Martin, and Kenneth J. Fasching-Varner. 2018. "Mapping the Margins and Searching for Higher Ground: Examining the Marginalisation of Black Female Graduate Students at PWIs." *Gender and Education* 30 (3): 295.

hooks, bell. 1989. *Talking Back: Thinking Feminist, Thinking Black*. Boston: South End Press.

Kanter, Rosabeth Moss. 1977. *Men and Women of the Corporation*. New York: Basic Books.

Kynard, Carmen. 2016. "This Bridge: The BlackFeministCompositionist's Guide to the Colonial and Imperial Violence of Schooling Today." *Feminist Teacher* 26 (2–3): 126–141. http://doi.org/10.5406/femteacher.26.2-3.0126.

Niemann, Yolanda Flores. 2020. "The Social Ecology of Tokenism in Higher Education Institutions." In *Presumed Incompetent II: Race, Class, Power, and Resistance of Women in Academia*, edited by Yolanda Flores Niemann, Gabriella Gutierrez y Muhs, and Carmen G. González, 325–331. Logan: Utah State University Press. https://muse.jhu.edu/book/74129.

Patton, Stacey. 2012. "The Ph.D. Now Comes with Food Stamps." *The Chronicle of Higher Education*, May 6, 2012. https://www.chronicle.com/article/From-Graduate-School-to/131795.

Perryman-Clark, Staci Maree. 2009. "Black Female Intellectuals in the Academy: Inventing the Rhetoric and Composition Special Topics Course." *Composition Forum* 20: 2.

Robinson, Subrina J. 2013. "Spoketokenism: Black Women Talking back about Graduate School Experiences." *Race, Ethnicity and Education* 16 (2): 155–181. http://doi.org/10.1080/13613324.2011.645567.

Rogers, Sean, Adrienne E. Eaton, and Paula B. Voos. 2014. "Effects of Unionization on Graduate Student Employees: Faculty-Student Relations, Academic Freedom, and Pay." *ILR Review* 66 (2): 487–510.

Shavers, Marjorie C., and James L. Moore. 2019. "The Perpetual Outsider: Voices of Black Women Pursuing Doctoral Degrees at Predominantly White Institutions." *Journal of Multicultural Counseling and Development* 47 (4): 210–226. http://doi.org/10.1002/jmcd.12154.

Smith, Leonie. 2005. "To Be Black, Female, and Literate: A Personal Journey in Education and Alienation." In *Readers of the Quilt: Essays on Being Black, Female, and Literate*, edited by Joanne Kilgour Dowdy, 183–200. Cresskill, NJ: Hampton Press.

Vo, Lind Trinh. 2012. "Navigating the Academic Terrain: The Racial and Gender Politics of Elusive Belonging." In *Presumed Incompetent: The Intersections of Race and Class for Women in Academia*, edited by Gabriella Gutiérrez y Muhs, Yolanda Flores Niemann, Carmen G. González, and Angela P. Harris, 93–109. Logan: Utah State University Press.

Wollast, Robin, Gentiane Boudrenghien, Nicolas Van der Linden, Benoît Galand, Nathalie Roland, Christelle Devos, Mikaël De Clercq, Olivier Klein, Assaad Azzi, and Mariane Frenay. 2018. "Who Are the Doctoral Students Who Drop Out? Factors Associated with the Rate of Doctoral Degree Completion in Universities." *International Journal of Higher Education* 7 (4): 143–156.

Young, Vershawn Ashanti. 2004. "Your Average Nigga." *College Composition and Communication* 55 (4): 693. http://doi.org/10.2307/4140667.

9

Literacy and Disciplinarity

Vignettes of Struggle and Identification

RAYMOND D. ROSAS

Introduction

Family lore has it that issues of literacy contributed to my great-grandmother's loss of a finger. Back at the beginning of the twentieth century, Victoria Gastelo was working as a servant for a wealthy Spanish family in southern Arizona. Spanish rancheros were common in the area up until about the 1930s, and they often relied on Mexican and Indigenous labor as traditional cattle ranching transformed into large-scale commercial production. The combination of her troubles with English—written and spoken—and her inability to cook a proper tortilla set the stage for violence. While my great-grandmother did indeed speak Spanish, I'm told she did not speak the *right* kind of Spanish. Her idiolect drew together elements of both Spanish and Navajo, which was apparently too much for her Spanish employer. One day, while her hands were still dusty and sticky from working the flour, salt, and lard that go into traditional Sonoran-style tortillas, her Spanish employer reached the limits of his patience with her cooking and code-meshing, found a hatchet, and proceeded to cleave my great-grandmother's ring finger clear off. I'm reminded of this incident every time I hold one of the few remaining images of her. In one photograph, she wears a black shawl and headdress proudly and resolutely. I'm

https://doi.org/10.7330/9781646426331.c009

told it was her usual attire for Sunday mass. A rosary is threaded through her hands; tracing the contours reveals a curious anatomical absence.

I open with this troubling vignette neither to sensationalize nor to shock. Rather, I relate this story to ground discussion of English literacy in a concrete example of the way language ideology can sanction violence, bodily or otherwise. Ultimately, my great-grandmother's code-meshing signified a deviation from prevailing language norms. And though the capacity of language beliefs and policies to legitimize harm is well documented, new instantiations continue to arise. Entrenched beliefs that connect language purity to national sovereignty and identity, for example, ignite calls for so-called English-only legislation and education policy. From *Lau v. Nichols* (1974) to Arizona's English for the Children initiative, debates about what to teach as well as where to place and how to fund English language learners (ELLs) often eschew questions of curricular innovation and the affordances of language variation for assimilationist fare regarding "national public language" and moral obligations to equip students, "regardless of their ethnicity or national origins, with the skills necessary to become productive members of our society" (Arizona Proposition 203, 2000). Of course, such warrants ignore research that finds modern work environments to be deeply plurilingual (Canagarajah 2013; Higgins 2009). They also ignore the generative capacities of language varieties—like African American Vernacular English (AAVE)—to help students find success in tertiary education and beyond (Gilyard and Richardson 2015).

My great-grandmother was still early in her experiments with English when the hatchet incident occurred. Although I can never be certain, as an educator I have to assume that the act fundamentally altered her relationship with language-learning. I have to assume that violence as well as the threat of violence obliterated any friendly feelings—if there were any in the first place—toward the language(s) of her employer. I am tempted to draw parallels between the scene above and contemporary measures to eradicate language variety in primary, secondary, and tertiary classrooms. I know, however, that I need to tread carefully. Though perhaps connected at the level of ideology, language policy and racist acts of bodily harm and violation are separate phenomena, each calling for their own strategies of deliberative and resistive engagement. Suffice it to say that both domains seek to foreclose possibility through fear: the fear of being an Indigenous woman in a Spanish, masculine space; the fear of producing a nonstandard utterance in sites of formal literacy instruction, wherein linguistic deviation is still too often associated with a myriad of deficits. As Karen R. Tellez-Trujillo's chapter in this volume so clearly illustrates, "all too often the adversity we face is in relationship to language, reading, and writing." Still, there is room in writing studies for hope.

At least two factors undergird this claim. First, the field's classical roots, wherein rhetoric was considered without subject but, nevertheless, essential for inquiry and argumentation on a variety of topics, promotes a level of capaciousness with regard to subject and method of inquiry. "Rhetoric," according to Cheryl Glenn, has always been "a plastic art" (2018). Second, ambiguity concerning the field's primary object of analysis might also constitute an epistemic advantage.

Understanding what writing is, how it functions, where it travels, and who wields its perlocutionary power may, according to John Mays, "be hopelessly indeterminate" (2017, 563). Indeterminacy, nevertheless, can be productive, especially where issues of language, identity, and positionality—key variables on the road to claiming a discipline—intersect in the idiosyncratic process of shaping personal affinities with disciplinary knowledge into, however tentative and subject to flux, disciplinary identity.

Writing studies has an established history of connecting the dots between lived experience and constructions of scholarly selves. Keith Gilyard's *Voices of the Self: A Study of Language Competence*, Victor Villanueva's (1993) *Bootstraps: From an American Academic of Color*, Mike Rose's *Lives on the Boundary: The Struggles and Achievements of America's Underprepared*, and Elaine Richardson's (2013) *PHD to Ph.D.: How Education Saved My Life* explicate the ways in which gender, race, class, and language competence interpolate graduate students into categories defined a priori. These writers demonstrate the generative nature of straddling the line between community and institutional literacy, the creativity that manifests when the literacy practices of "the street" catalyze something new in more formal, at times hostile, sites of literacy education (Gilyard 1991, 160). Furthermore, they highlight where sites of literacy instruction could do more to promote equity across activities of textual production, consumption, and evaluation. Interrogating the logics behind these core activities is of critical importance to graduate students in writing studies, as each one has served as a vector of much politics at one point or another. As the late Mike Rose observed, "nothing is more exclusive than the academic club: its language is highbrow, it has fancy badges, and it worships tradition. It limits itself to a few participants who prefer to talk to each other" (1989, 58). On the road to disciplinary identification, then, graduate students are interpolated into balancing acts, whereby lived experience and disciplinary doxa interanimate and struggle for prominence.

Writing studies fosters this type of balancing act—a disciplinary feature that I find especially advantageous. In what follows, I give voice to several vignettes in order to highlight key moments on my road to disciplinary identification. By sharing these vignettes, I hope to illustrate the flexibility of writing studies in terms of accommodating research interests sparked by lived experiences. The organization of this brief chapter follows three themes of identification, each one punctuated by the aforementioned concerns of literacy, identity, and violence: (1) claiming the discipline as a classed subject from a working-class barrio, (2) claiming the discipline as a person who experienced the ins and outs of the US opiate epidemic, and (3) claiming the discipline as a raced subject with a critical approach to academe's role in fostering equity. In my view, to claim writing-studies-as-discipline is to assume responsibility for illuminating the potential of equitable literacy practices and for mitigating the effects of vernacular and discursive hegemony across contexts of literacy education.

Pivot 1: Early Encounters with Literacy Myths and Respectability Politics

I'm trying to write my name and I'm struggling. I'm not trying to overachieve or anything like that, so I'll go with the shortened version of my name: three simple letters. Still, I can't seem to get the pencil to work. And I need to write my name in two places, once on the board and once on the construction paper I'm staring down at. R is a crazy letter to compose for the first time; so is Y. I think if I just wait long enough, the letters are bound to emerge. They don't. Now I'm embarrassed; now tears are streaming; and now I'm wishing I was home.

Kindergarten set the scene for my earliest memory of the writing process. I remember being absolutely mesmerized by the ease with which some of the other students jotted their name; I also remember what that did for my writerly self-concept. From kindergarten forward, I felt a need to prove myself via the written word. At times, my eagerness to engage school literacy got me noticed for the wrong reasons.

I didn't have language for it at the time, but Mr. Martinez, my fourth-grade Language Arts teacher, loved getting into respectability politics. He was always telling us to "stop wearing those lowrider shirts" to class and to "stop using so much tres flores in your hair," which is a hair gel popularized by Chicanx youth. Apparently, too much hair grease and the slicked back fade made us look like "hoodlums." Still, we could tell that Mr. Martinez cared. He had a gift for feedback, the kind that was at once pointed, uplifting, and instructive. In terms of pedagogy, it was as if he was preparing us for some far-off academic battle. His operating ethos seemed to be the following: *People are going to read your writing and pass judgments on you and, in turn, your family. Thus, our mission is to do everything we can to prevent those judgments from sliding into the pejorative.* I had no idea when, where, or if this battle was ever going to materialize, as nobody my age in the neighborhood lost much sleep over the finer points of writing (at least I thought so at the time). Most of us, however, took his sermonizing to heart.

That's why I was proud of myself when I completed three essays way ahead of schedule. I had read ahead and wrote ahead and expected to be rewarded for the productivity. Eager to turn the essays in—and because my parents didn't own a stapler or have paperclips handy—I simply tore three pieces of masking tape, placed them at the upper-left corners of the essays, and handed the work in:

> ¡Mira! You did good hijo! All three of them done already? But what's this? ¿Qué es esto? Why is there masking tape here? Making a piñata? What, you couldn't go out of your way to find a stapler? This isn't Mexico, señor! This is not Mexico. No sir. Go to the library and find a stapler!

I was mortified. I had sidestepped convention and paid the price, which consisted of a lonely walk to the library in ninety-degree heat. Head down. Despite my efforts at composing well-written arguments, the materiality of the essays got me in trouble. This, remember, was supposed to be a triumphant day, one where I cashed in on my literate efforts. The message was clear: use staples, never tape. In the battle for respectability, I had just surrendered my ethos.

You can read this incident in a variety of ways. To start, the quip about Mexico calls to mind the vexed relationship between space (location) and literacy. For Martinez, the materiality of the drafts became emblematic of a different writing ecology, one at odds with the expectations of schooling in the United States. Tape indexed a writing system where the common implements of academic writing—staples, paperclips, etc.—could not be taken for granted. The issue was not framed as one of assignment requirements, aesthetics, or even conventions; rather, the issue at hand is about place or, perhaps more accurately, the denial of place: "This is not Mexico," Mr. Martinez argued. By doing so, he bordered the rhetorical possibility of the classroom—and this was despite all the code-switching. Classrooms are, of course, circumscribed by ideology like any other social arrangement beyond academe. Numerous stakeholders infuse spaces of literacy with political and ideological discourse, which, in turn, introduces hierarchy and stratification into literacy instruction.

Again, Mr. Martinez cared a lot about his students, and the point here is not to construct a diatribe against his pedagogy. But on the road to disciplinary identification, this moment was pivotal. I knew at the time that this was an interesting moment, though I wouldn't have the language to analyze it until about twenty years later. Here was a moment overdetermined by attitudes about nationalism, ethnicity, linguistic/discursive practice, and the materiality of writing; here was a moment that left me, as a writer, shook. Shook over norms I didn't even know existed; norms that were neither connected to the substance of the arguments I was putting down nor communicated explicitly. Such foci draw considerable attention from scholars in writing studies. I am reminded here of Shirley Heath's (1983) classic contention that "the language socialization process in all its complexity is more powerful than such single-factor explanations" like formal correctness (344). I had completed what was asked of me and still missed the mark. I needed answers to my remaining questions about the whole masking-tape-Mexico-writing episode. I don't mean to frame this experience as epiphanic. I was a kid, and the politics of writing instruction did not hold my attention for long. However, from a disciplinary lens I can see how this moment could have been pivotal for a myriad of reasons, and I often reflect on this moment as a means of scholarly invention. Indeed, I consider this critical incident an integral thread in my claiming-the-discipline narrative. In this episode on my road to disciplinary identification, I gleaned that tropes of literacy are complex and problematic but, nevertheless, generative problem spaces for exploring how writing studies frames the promise of our work.

Pivot 2: Positional Resources for Scholarly Invention

I was getting tired of sleeping in my truck in one-hundred-degree-plus weather. When you are up to your neck in addiction, you have to be ready to travel to all parts of the city to keep up the habit. This is not a New York City smack story; this is urban sprawl in the desert; you need to stay glued to your car to optimize chances of scoring. Heroin dictates where you go, when, and with whom; throw questions of agency out the window.

On this day, I'm hauled along the highway to a familiar north-side scene to partake in an increasingly familiar sensation. As I pull into the apartment complex, I immediately notice the string of cops parked in the vicinity of my connection's apartment. Now, any sensible person would stop, shift to reverse, and slowly and carefully drive away—the vehicular equivalent of tiptoeing. Not me. I had to know whether my tether to the world was about to be cut. So, I drive on, slowly and confidently. Three cars in a row. In the first, nothing; in the second, still fine; in the third, the outline of a person with their head down, shoulders pulled way back because of the way "hands behind your back" externally rotates them. It was confirmed. And before the thought of coming down crossed my mind, I was already strategizing for the next hit—stopping was out of the question. New settings had to be explored.

South Tenth was always considered the worst of the worst of all the south-side avenues. At least South Tenth after a certain hour. I had never believed the rumors, however. The avenue showcases some of the best Sonoran Mexican food in the area. It's an olfactory heaven. The aroma of freshly baked bolillos, tortillas, and pan dulce intermingle with the smells of various mesquite carne asadas. On Sundays, families come from all over the area to purchase menudo by the bucket full. It's a place of family; a place of Sunday slow jams and raspados. That's the South Tenth avenue I knew, and that's one I visit today. Of course, like all barrios, there is another side.

Driven by necessity, I decided to see if the rumors about South Tenth avenue were true. There is a sharp contrast between the linguistic landscape of the suburb and that of the avenue. Bilingual advertisements dominate the discursive real estate of the area, a feature that added to the sinking feeling developing in my gut. In short, encountering blends of English and Spanish hit close to home. I had been consistently getting my fix in the white suburbs of the northside, but now I was only a ten-minute drive from the neighborhood of my childhood. That realization hurt, but the hurt could wait—and it could always be nullified.

I'm naïvely driving up and down the avenue when I'm waved down by a woman in a long wool coat. She must have noticed how the desperation of withdrawal was contorting my face:

¿Qué quieres? Are you a cop? Prove to me you're not a cop. Show me your belt then. Okay. Todo bien. Todo bien. A cop wouldn't have let me do that. I can hook you up down the street. You got cash, right? Good. We can both get some. I'm dope sick too, mi hijo.

Code-switching helped me communicate with fellow addicts to find heroin when the suburbs dried up of prescription opioids. When the suburbs were wiped out of their supply, I could talk my way into the most exclusive street scenes and trap houses. I hadn't the slightest clue that linguistic repertoires and language variety were considered interesting language phenomena; I was simply shuffling from one discourse community to another in order to keep withdrawal at bay.

I was a person of color in what is widely portrayed as a white phenomenon. The public health tragedy we know as the opiate epidemic has engendered significant debate about the role of Big Pharma in addicting entire communities to prescription pain killers. Furthermore, that we invoke the authority of medical discourse—*epidemic* as opposed to a war on drugs and associated users—says something about the ways in which race is made salient in the contemporary public imaginary. In a crisis with white suburbia at the epicenter, the medical model supplants the criminal one of previous public health disasters regarding narcotic drug consumption (Netherland and Hansen 2016). Stuart Hall (2019) reminds us that when considering historical transformations of race and racism,

> the question is not whether men-in-general [sic] make perceptual distinctions between groups with different racial or ethnic characteristics, but rather, what are the specific conditions which make this form of distinction socially pertinent, historically active? (58)

Hall's insight constitutes a major through line in most of my work. My current theoretical commitments map onto what was easily the darkest period of my life via my work examining the rhetoricality of the opioid epidemic and its constructions of white, sympathetic users (as opposed to the minoritized criminals of past drug scares). Researching and writing about the wider context of my personal experience with addiction has been exceptionally cathartic. But more to the point of this edited collection, writing about the opioid epidemic has provided me a means to claim writing studies in ways that feel authentic.

In short, I claim and identify with writing studies because, at its best, the field embraces capacious epistemologies for explicating social and linguistic phenomena that I have myself experienced. The tragedy of the opioid crisis cuts across all identity markers, and the opening critique is not meant to deny the suffering it fosters across the United States and beyond. I simply mean to point out that, from a rhetorical perspective, the construction of medical subjects (when considering white bodies) versus the construction of criminals (when considering minority bodies) demands explication and critique. Part of that construction is no doubt related to social structure, and part of it is related to discourse. My point, however, is that writing studies makes room for using lived experience as a catalyst for not only disciplinary change but also for a means of stimulating change in the public sphere. No doubt this may sound naïve. However, as a Cheryl Glenn advises, I'm compelled to buy in to the generative capacities of "this thing called hope." Hope got me through my addiction, and I'm hopeful that writing studies scholars can contribute to meaningful social change across a range of issues. In this episode on my ride to disciplinary identification, I gleaned that mining lived experience was an integral rhetorical tactic for not only scholarly invention but also for claiming writing studies.

Pivot 3: Embracing and Contesting Disciplinary Subjectivities

The fall semester is ending, deadlines are fast approaching, and I'm going to be writing for more than a minute. Despite telling myself repeatedly that I would complete all my writing projects way ahead of schedule, I find myself in the familiar position of being the last person in my section of graduate student cubicles. The university has just announced early closure due to a winter storm. I'm in a positive frame of mind because to somebody like me from the Southwest, snow is an exciting—even thrilling—meteorological event. The thrill is fading, of course, the longer I stay in the Mid-Atlantic. But on this occasion, I welcome the snow as a pleasant backdrop to my idiosyncratic composing process. My writing process is also kinetic, which is to say that I often pace when working through an idea. On this day, I'm pacing the hallway just outside my cubicle.

Mid-pace, I recognize a staffer that I often see around the department. I say "hello" to be friendly, expecting to generate the standard adjacency pair of "Hi, how are you?" "I'm fine, thank you. How are you?" However, what I get is completely unexpected. In response to my "hello," I'm met with "are you supposed to be here?"

The staffer likely had no idea that they just gave voice to a thought that punctuates my experience of academe. I felt shocked and humiliated all at once. I had read about things like this happening to other academics of color, had read about the so-called imposter syndrome, but I hadn't yet experienced such an explicit incident. What about me suggested that I didn't belong in this space? I went through a whole list of possibilities and justifications to make sense of why my positionality invited such a response. I knew, of course, what it was all about. Yet when racism manifests in dyadic communication, it has a way of catching you off guard. "Tussles for inclusion," argues D'Angelo Bridges, can occur at any time and any place in academe (2021, 231).

Subjectivity becomes especially salient in contexts of higher education. Beyond its well-known implications for educational attainment, the inventional capacity of positionality is susceptible to co-optation. In higher education, the commodification of people of color—as subjects of analysis, as appointed guardians of diversity and inclusion, as tokens—is an issue that reflects the staying power of structural conditions designed to delimit and proscribe minority achievement (Baez 2000). To concretize, consider Ersula Ore's (2017) experiences as a woman of color on the English faculty in a predominantly white institution (PWI). As result of her positionality as a woman of color, Ore was subject to questions about whether she belonged in and whether experiences circumscribed by questions of whether she was in the right place or had the professional authority to walk the halls in the humanities building. Ore argues,

> the university is not a space detached from the world outside. America's long history of racial power, heteropatriarchy, and classism saturate the social space beyond the university just as much as they saturate the space within it. (2017, 18)

In considering Ore's characterization of academic space, the literacy-as-uplift trope looks increasingly untenable for certain minoritized populations. To date, the payoffs of literacy remained mixed, with scholars marshaling evidence on either side of the literacy-as-myth, literacy-as-uplift divide. My own position on the matter is still forming, but I will say I'm attracted to recent work that complicates Harvey J. Graff's (1979) explication of the relationship between literacy and cultural hegemony (see Baker 2014). And, to harken back to the vignette of Mr. Martinez and his respectability politics, the notion that literacy equates to social uplift still motivates numerous language arts pedagogies and other philanthropic efforts. (For a germane example, check out

UNESCO's rhetoric of literacy. According to this organization, literacy is key to empowering communities in terms of social mobility and general well-being). Thus, I'm hesitant to completely part with literacy myths; as we all know as analysts of discourse, myths are powerful motivational forces even if their empirical basis is at best tenuous.

I also maintain a modicum of hope for the work ahead, the type of work Glenn (2018) contends will lead us to a disciplinary location where we may "write, speak, and teach the words that reshape (and repair) the world and pave our future" (212). In this episode on my road to disciplinary identification, I gleaned that crafting a space of hope for our work requires openness and flexibility. This is a requirement that aligns perfectly with the plasticity and epistemic core of this thing called writing studies.

Conclusion

From helping beginning writers find meaningful expression to contesting hegemonic rhetorical practices across various publics, the possibilities of this field are radiating and multiplying as new generations of writing studies scholars write themselves into the discipline. Composing one's scholarly self will always be a balancing act, especially for BIPOC scholars and researchers who must straddle the at once real and imagined line between home and school linguistic culture, as well as the line between inclusion and tokenism. As Khadeidra Billingsley reminds us in her chapter in this volume, "Tokenism is a laborious experience." The rhetorical conditions that inform a graduate student's decision to claim writing studies are no doubt varied and multiple, as unique as the idiosyncrasies of our prose. However, and at the risk of sounding redundant, this field is plastic and therefore accommodating of the myriad ways one might strategically claim writing studies.

For me, claiming writing studies will always be a gesture to the memory of my great-grandmother and her violent encounter at the nexus of language ideology and positionality. This is how I balance my familial allegiances and my academic ones. This is how I remind myself of the obligations I have to the communities that nourished my intellectual curiosity in the first place. In short, keeping the memory of my great-grandmother's narrative alive is a rhetorical tactic I deploy to claim writing studies as my home discipline.

In my scholarship and pedagogy, I strive to foster critical literacy, which is "both a narrative for agency as well as a referent for critique" (Gilyard 1991, 28). In marshaling this strategy, I lay claim to writing studies in the knowledge that literacy—in all its varieties—is at once a vehicle for freedom and harm, a vector for conflict—its generation, perpetuation, and, hopefully, resolution. Literacy is a generative problem space, and writing studies provides me with the tools to disambiguate what future claims on literacy might look like. At this moment in my graduate student career, I believe claiming writing studies is about constructing equitable futures through the critical explication of discourse; I feel the full import of such a mission when I reflect on the narrative of my great-grandmother.

–Con Safos (Chicanx phrase that signifies "with respect, with safety")

References

Arizona Proposition 203. 2000. Arizona Secretary of State. 2000 Ballot Propositions. https://apps.azsos.gov/election/2000/Info/pubpamphlet/english/prop203.htm.

Baez, B. 2000. "Agency, Structure, and Power: An Inquiry into Racism and Resistance for Education." *Studies in Philosophy and Education* 19 (4): 329–348. https://doi.org/10.1023/A:1005241732091.

Baker, David P. 2014. *The Schooled Society: The Educational Transformation of Global Culture*. Stanford, CA: Stanford University Press.

Bridges, D'Angelo A. 2021. "Sibling-Scholar Network as a Means of Survival." *Rhetoric Review* 40: 207–256.

Canagarajah, A. Suresh. 2013. *Translingual Practice: Global Englishes and Xosmopolitan Relations*. New York: Routledge.

Gilyard, Keith. 1991. *Voices of the Self: A Study of Language Competence*. Detroit, MI: Wayne State University Press.

Gilyard, Keith, and Elaine Richardson. 2015. "Students' Rights to Possibility: Basic Writing and African American Rhetoric." In *Insurrections: Approaches to Resistance in Composition Studies*, edited by Andrea Greenbaum, 37–51. Albany, NY: SUNY Press.

Glenn, Cheryl. 2018. *Rhetorical Feminism and This Thing Called Hope*. Carbondale: Southern Illinois University Press.

Graff, Harvey J. 1979. *The Literacy Myth: Literacy and Social Structure in the Nineteenth-Century City*. New York: Academic Press.

Hall, Stuart. 2019. *Essential Essays*, 2 vols., edited by David Morley. Durham, NC: Duke University Press.

Heath, Shirley B. 1983. *Ways with Words: Language, Life, and Work in Communities and Classrooms*. New York: Cambridge University Press.

Higgins, Christina. 2009. *English as a Local Language: Post-Colonial Identities and Multilingual Practices*. Bristol, UK: Multilingual Matters.

Mays, Chris. 2017. "Writing Complexity, One Stability at a Time: Teaching Complexity as a Complex System." *College Composition and Communication* 68: 559–585.

Netherland, Julie, and Helena B. Hansen. 2016. "The War on Drugs That Wasn't: Wasted Whiteness, Dirty Doctors, and Race in Media Coverage of Prescription Opioid Misuse." *Culture, Medicine and Psychiatry* 40 (4): 664–686.

Ore, Ersula. 2017. "Pushback: A Pedagogy of Care." *Pedagogy* 17 (1): 9–33.

Richardson, Elaine. 2013. *PHD to Ph.D.: How Education Saved My Life*. Philadelphia, PA: Parlor Press.

Rose, Mike. 1989. *Lives on the Boundary: A Moving Account of the Struggles and Achievements of America's Educationally Underprepared*. New York: Penguin Books.

Villanueva, Victor. 1993. *Bootstraps: From an American Academic of Color*. Urbana, IL: National Council of Teachers of English.

10

Cognitive Dissonance

ALISON WELLS ZEPEDA

In the fall of 2018, I sat in the university writing center at my institution and eagerly listened to Anne von Petersdorff, PhD, a scholar in German studies and digital humanities, detail the documentary that became her dissertation. Her autoethnographic work enviably took her across several different countries as she experienced unfamiliar cultures and living conditions, showcasing female intersubjectivity and embodiment produced through the lens of the camera. von Petersdorff (2018) advocates "experiencing the fringes" in scholarly work and argues for a cross-disciplinarity that draws strongly on one's positionality and the incorporation of hybridity, as evidenced by her project. She further discussed the importance of making and keeping one's scholarly work personal through focusing on one's original contribution to knowledge; in effect, for von Petersdorff this is central to how knowledge is created. My own concept of knowledge is defined by the observation and processing of linguistic, material collaborations and contributions between and among human and nonhuman bodies in the environment. Knowledge is therefore always becoming, unfolding, and defined by impermanence. What has stayed with me most, though, is the idea that one's writing is defined to a great extent by one's material, temporal, and relational limits. While I admire scholars who have the freedom and resources to globe-trek for the purposes of academic travel and research projects, I have no choice but to acknowledge the extent of my personal limitations and my historic reliance on language, which has shaped my personal and academic life as well as afforded me a scholastic perspective that has both challenged and hindered me. My claim and early theoretical work in the discipline of rhetoric and writing studies is a culmination of how I have understood the tangled relationship between the world, myself, and all others. It is an exercise in exploring the relationships between the material and feelings; in effect, how "what's lying around in the environment" (Rickert 2013) produces affect and makes space for rhetoric—without language. I am intrigued by the concept of the posthuman, which has enabled me to move beyond language as my primary vehicle for knowledge and understanding. I am especially concerned with the inability of the human to trust intuition, to let go of the structuring capability of language to construct one's mind.

Wanting Words

From a very young age, I have been enchanted by words and, consequently, expected a great deal from them. Like many others, I have been captivated by the idea that language, and my ability to use it correctly and stylistically, serves as the primary vehicle to academic success and, from there, all other successes in life. In an early childhood development center at the university where my father worked, I used Mr. Sketch markers to carefully scribe words I already knew how to spell on stiff, bookmark-shaped cardstock bound with a metal ring. I added to the ring and reviewed—what I already knew—over and over until I felt I had mastered those words. A few years later, I was gifted the 1962 edition of the *World Book* and read voraciously, often staying up past bedtime to satisfy my desire to know through words. My family recognized this about me and furnished me with novels that were generally outside of my interest and attention span (later I was always given blank journals—all of which have remained mostly blank—because it was presumed that I liked writing in my free time because I enjoyed reading so much). They assumed that because I was a proficient reader that I understood all that I read or that I enjoyed encountering words I didn't know. I preferred to stick with that which was familiar: I read and reread young adult fiction that resonated with me. Nothing challenging. I felt I could read well but was not always able to comprehend. I realize now that saying that I can read is incongruent with saying I can read that which troubles or challenges me or my dominant position over words that I knew. The words I love and were familiar with never changed, the stories never changed, and my feelings about them never changed. Reading the same books over and over provided a stability like no other; I also see now that I have always been searching for solid ground on which to stand, on which to base my life. Now I understand that the desire for stability in an uncertain world is simply one part of what it means to be human. Now I realize that there exists an inarticulable, affective element that not only diminishes language and rhetoric as single handedly responsible for all that humans know, but underscores all human and nonhuman interaction as a priori.

My claim to the discipline of rhetoric and writing studies is fundamentally informed by a longstanding desire to question what we value and teach in rhetoric and writing courses and how we go about teaching those shifting, temporal values that are seemingly impossible to find, to pin down once and for all (Vitanza 2003). In carefully considering Vitanza's assertion that writing simply wants—not what humans want writing to do, not what humans want from writing, and not what humans gain from writing—we are forced to interrogate our relationship with writing and the institutions that govern writing pedagogy as one that is focused primarily on a product marked by stability and correctness. As such, a posthuman writing disposition is guided by the notion that writing cannot produce exactly what humans want because it, as a practice, is granted ontological equity with humans and all other bodies, human and nonhuman. Posthumanism is simply not about what humans want from any other body, but it ethically requires us to ask what is wanted by human and nonhuman others.

Ecological Complexities

My interest in theories of the posthuman, therefore, is driven by a preoccupation with the notion that humans do not preside in positions over all other forms of matter including writing and its deliverables, which implicates us to acknowledge the "muddy, messy conditions of our existence" (Coole and Frost 2013) with regard to the ecological, environmental, political, and technological—material—entanglements we find ourselves in. In other words, the world and our positions in it have been made theoretically foreign under posthumanism, as humans have historically ascertained our unyielding power over things, people, and events we believed we had control over. It is this human acknowledgment and consequent relinquishing of power that compels me to rearticulate for myself and for those I teach and learn with how this ontologically complex disposition toward existence changes how we and our students write with, around, about, in response to, and in the world. This is especially true in light of the complexities in responding to, for example, the gravity of human and nonhuman ecological concerns regarding climate change, polarizing human ideologies and realities driven by political rhetoric and discourse and defined by an uninhibited persuasion in which nothing is off limits, and biological threats that we cannot foresee—all examples of the invisibility that characterizes the affectual complexities and constant becoming of material and bodily interactions, human and nonhuman.

INTRADISCIPLINARY CONFLICT

A recent, ongoing debate among contemporary rhetoric and writing studies scholars involving feminist assertions of rhetoric's materiality and proponents of object-oriented ontology has illuminated the juncture in rhetorical studies that prompts my disciplinary claim and research. Historically, feminism has been concerned with accounting for power imbalances between bodies, a "volatile" undertaking that has relegated the feminisms to "taking refuge within culture, discourse, and language"; however, Alaimo and Hekman (2008) suggest that "feminist theory is at an impasse caused by the contemporary linguistic turn in feminist thought" (1). Dismantling power imbalance(s) between bodies has been, historically, the central focus of feminism. While language, defined here as a strictly human, material expression that reveals an intrinsic human need for communication, has wielded its power in favor of producing feminist theory and laid the foundation necessary for making known the "pernicious logic that casts woman as subordinated, inferior, a mirror of the same, or all but invisible," it has detracted attention from the body's ecological materiality, which reveals a corporeal reality more than one painted by words and discourse *about* the body (Alaimo and Hekman 2008, 2). Material feminism is interested in exploring agential relationality with all forms of matter, human and nonhuman, and how matter acts both alongside, because of, and independently of language and discourse. Recognition of the material conditions of my own life compel me to acknowledge that my social identity is shaped by my body and language surrounding it, as well as its role in how my own reality has been shaped through my body's material relationships. In exploring the concept of the posthuman and the externality that it depends on as a theoretical concept, the body as relevant to feminism is a central facet to the concept of the posthuman.

I am similarly and, perhaps, contradictorily, drawn to object-oriented ontology, a set of theories ostensibly supported by white men in the discipline of rhetoric and writing studies. It seems to account more effectively for the rhetoricity of matter—that is, how matter is agential—and is not focused solely on human relationality with other forms of matter. The openness and "flat ontography" (Zabrowski, 2016) supplied by object-oriented ontology attends specifically to the delegitimizing of human exceptionalism so that all forms of matter can be granted equal rhetoricity, and therefore, agency. Based on this premise, it could be said that object-oriented ontology acts as a form of whiteness, reinforcing and enacting a kind of sameness as difference model for examining the rhetoricity and relationships of bodies, both human and nonhuman. By displacing the human as central to knowing and depriveleging human thought as superior to other forms of logic, object-oriented ontology has the grave potential to remove the lived, material experiences of people as simply being part of a network of matter, potentially stripping people of identity as constructed through human social interaction. Human social construction, including all forms of communication, both material and immaterial, is responsible for maintaining humanity as the ultimate position to such a degree that we have used words to simultaneously dehumanize and exoticize the Other to preserve perceived power and positions as supreme above all else (Said 1978). To erase language or diminish its ultra-significant role from human experience would be to erase what it means to be human; however, by placing language as material on the same plain as all other forms of matter, it's possible to acknowledge language as having no greater agency than any other material, human or nonhuman.

RHETORIC'S MATERIALITY: A CLASH

Recent scholarly movements in the discipline illuminate the "significant challenge" rhetoric faces in, at the very least, incorporating objects into the sphere of what counts as rhetorical, recognizing "the way things are and the rhetorical force they wield in relation to us and other things" (Barnett and Boyle 2017, 2). The cumulative relational, interactional nature of all things, materials, and bodies produces affect characterized by a flood of material interactions both human and nonhuman, which produces a form of opaque knowing—very much like a photograph featuring a subject and a backdrop that made the photograph possible. It's true that the backdrop isn't necessarily always considered, rather, it is the occasion that is remembered *because* of the backdrop. A rhetorical, affectual energy is drawn from the presence and composition of the existing arrangement of materials, bodies, and objects. Thus, my scholarship interests are rooted firmly in the affective rhetorical spaces that exist between humans and nonhumans. My positionality guides and complicates my research and writing in such a way that I must acknowledge how my theoretical disposition both informs and muddles my thinking and is responsible for that which I have been socially groomed to uphold as a white woman. Because of my attentiveness and orientation to language and words, while also being keenly aware of how I am affected by the *things* in my environment—photographs of my grandparents that rest on the desk my father built me as I write, an abstract statue of a ballerina embodying the memory of an art I have loved and practiced, a grammar textbook published in 1913 that I inherited from my grandmother—that I am interested in how we know what we know in the absence of language.

The discipline of rhetoric and writing studies has provided the space for me to explore the clash between language and material. Human language cannot exist independently of the material, objects, and all bodies, making the human completely dependent on what exists, what we've created and reciprocally benefited from. My scholarly research of the posthuman, affect theory, and the convergences with rhetoric as energy have compelled me to not only recognize but fully acknowledge the openness required in this inquiry toward a challenging but necessary acknowledgment of new materialist work to ignore identity-based bodies—bodies that have been undeniably and often painfully shaped by what we have typically thought of as the social. In other words, it is inherently unethical to ignore that which has constructed our human bodies and have, indeed, played a role in our individual and collective experience and exposure to objects that exist both because of and in spite of humans.

Perhaps the human body, in its externalized complexity, is the most helpful metaphor for how to approach writing studies through the various lenses and angles of new materialism. While it is necessary to acknowledge that the human is not the locus of knowledge- and meaning-making through its most valuable, yet egregious, gift—language—it has clouded our collective view of ourselves, who we are, how we relate (or not) to some bodies and not others. Decentering language as the primary source of human interaction and relatedness compels us to find different ways of relating; it forces us to consider the entirety of the human as not only socially constructed but possessing the capacity to affect and be affected in ways that are often invisible.

Clary-Lemon (2019) argues for a thorough review of the whole of scholarship involving new materialist approaches in rhetoric and writing studies beyond the "cast of characters" currently circulating in the research strands of the posthuman, affect studies, new materialism, and object-oriented ontology. She recommends that scholars place Indigenous knowledge and scholars at the front of these studies in an effort to ethically resituate this thinking away from "primarily Western philosophical world views to the complete ignorance (often best case) and willful negation (worst case) of the work of Indigenous thinkers and writers" (Clary-Lemon 2019). These strands of thought have typically not given enough attention to the social construction of human bodies and the very real implications of those constructions in everyday life. Even Latour suggests that these concerns are best left to the sociologists, reinforcing the notion that the human body and its very real experiences can be flattened to the level of all other objects, materials, and their interactions (Clary-Lemon 2019). My theoretical work invites me to examine yet another strand of research that seeks to make sense of racial disparities—beyond that which social construction will have us consider.

My claim in the discipline of writing studies, thus, is a convergence of my proclivity for language and the nagging feeling that while language has been definitive of social construction and has created an academic space for me, it fails to account for the seemingly neutral, invisible, but affective and therefore agential spaces between all bodies and forms of matter, human and nonhuman. As a result, my disciplinary pursuits are firmly rooted in a paradoxical, persistent wrestling with my intrinsic need for intellectual and epistemological stability in a world that I have found in language, in conjunction with the realization that language is inherently inadequate to account for all forms of knowing—a realization that, for me, demands greater attunement to that which we owe our humanity or that which we have long defined as human. While this understanding feels like an original discovery for me, and certainly my scholarly work is deeply informed by my own experiences and observations about language and the role it has played in shaping my thinking about the theory and practice of writing and rhetoric, it is not new to the discipline (Worsham 2015; Hawhee 2017; Rickert 2013). A disciplinary focus in rhetoric and writing studies has provided me with the language I have felt I have needed to fully articulate my observations, which stands in clear dissonance to my need to simultaneously acknowledge affect as a necessary rhetorical partner without cementing language as the final word, the sole creator of knowledge, the foundation from which all is built. I have claimed the discipline through sheer uncertainty and trepidation of words and yet have found an absolute necessity to harness my theoretical observations through language I have had to learn through scholarly research and putting my own views in conversation with those who have shaped our historic understanding of rhetoric and its role in shaping reality.

RESISTANCE TO LANGUAGE AS TRUTH

I am contradictorily reliant on and resistant to language as a primary way to understand the material—human and nonhuman—interactions that construct our lives and existence, and this contradiction lies at the center of my claiming the discipline of writing studies. As a result, I have approached the discipline as an exercise in exploring the relationships between language, material, and affect. From the time I was a young white student in a historically Black elementary school, newly integrated, to the year I began graduate studies in technical communication, I have observed intently the complexities of how the environment shapes and is shaped by human and object interaction: the aged school building in the thick, humid woods, the long evenings writing and fumbling with new online technologies to do graduate work. Barad (2003) suggests that an overreliance on language has ultimately deterred us from considering material interactions not only as fundamental to life but primary in how those interactions produce affect—ultimately guiding our actions and thoughts in such a way that creates the need and desire for language. Language enables naming how and what we think, know, and feel; however, an epistemic dependency on language enables a creative construction that is often not representative of the "truth" of material interactions (Martin 1991) and takes on what we traditionally consider rhetoric—a version of the truth made possible and enhanced by language. This is especially evident in recent political media in which the public has become more acquainted with concepts that have become of interest to the public through political discourse like alternative facts and gaslighting. Language has been central to human knowledge-making, and the ability to read and write manifests in the everyday with every social media post, every impromptu scroll of the news feed, every text. While digital media use and technological advancements have made possible our need and perception of connection, the invisibility of those connections makes their origins opaque, at best. We are unable to grasp the interactional, often invisible, forces that create what we see in our screens. Our screens offer a looking-glass view of what we write (both literally and metaphorically) and is inevitably returned to us through our digital participation, yet the virality of rhetorical production through media is evidence that we are unable to detect the origins of posts, texts, ideas that are circulating based on an unfolding of events characterized by a collaboration among, between, because of, and in spite of an infinite collective of human and non-human actors.

RHETORIC'S POTENTIAL

In beginning coursework and tutoring students at the university writing center as part of the introduction to my graduate program, it became immediately evident to me that I would be able to acquire and employ the extensive language of the discipline of rhetoric and writing studies to tackle the academic questions that I had been unable to answer. During coursework, I found myself vacillating between thinking that everything is rhetoric to everything is merely rhetorical to imagining a world in which there is no social construction as we know it, and along with that, no digital technologies, I began to think more deeply about what counts as rhetoric. Most often, rhetoric has been conceptualized as a tool of language—the arrangement, composition, and delivery of words and images and the ways it persuades (Aristotle and Kennedy 1991) and shapes (Scott 1967; Brummett 1979) individual and collective reality. In this way, rhetoric is traditionally viewed as epistemic, producing what we know through language. I became very interested in rhetoric as a "form of mental energy" (Kennedy 1992). Prior to Kennedy's assertion, however, Aristotle's concepts of *energeia* (actuality) and *dynamis* (potentiality) "are conjoined ways of existence . . . with the sort of ambiguity that comes with its versatility" (Ingraham 2018). Energy as a rhetorical concept, because of its philosophical fluidity and (in spite of) "referential elusiveness," has generated an abundance of scholarly work in attempting to understand energy as "an inner force of sorts that can drive and impel work or activity" (Ingraham 2018). Based on Kennedy's notion that "the emotional energy that impels the speaker to speak, the physical energy expended in the utterance, the energy level encoded in the message, and the energy experienced by the recipient in decoding the message," it must be acknowledged that energy is fundamental to all affect and action, giving vitality to all forms of matter, human and nonhuman (Kennedy 1992, 2, as quoted in Ingraham, 2018). Much recent scholarship has followed Kennedy's lead in decentering the human as the sole producer of meaning and knowledge through language, spoken and written, and "place[ing] all things on a more level ontological footing, and to consider our shared problems more relationally and processually . . . often evoke 'energy' to do so" (Ingraham 2018). Rice (2005, as quoted in Ingraham 2018) refers to energy as rhetorical enacting "the viral intensities that are circulating in the social field . . . it is not the situation, but certain contagions and energy." Ingraham (2018) refers to Chaput's (2010, 263) recommendation that we "pay attention accordingly to the rhetorical energies surging throughout the concrete sites of our contemporary world," and offering, for me, a preliminary glimpse into what might comprise a posthuman writing curriculum. The challenge here, of course, is to reconceptualize writing and its visual, material manifestations as derived from "an affective dynamism that cannot be translated into language without loss, although that makes them no less rhetorical" (Ingraham 2018, 263). I am also brought back to one of my primary research questions: How do we know what we know without language helping to construct reality?

For a final paper in a rhetorical history course, I attempted to make the case for including an ancient American Southwest culture as a rhetorical tradition, even though there are no written records or verifiable language of the Mogollon people. The remains of the culture are composed of pottery shards scattered across New Mexico and Arizona with faint pictographs in cliff dwellings and other caves. These remnants of the culture suggest for me the centrality of material, human activity on technology and writing and the interconnectedness between the two. Technology is inherent in writing; writing is made possible by technologies—from the pencil to the word processor. Aristotle confirms this relationship as vital to rhetoric as "the available means." Therefore, one's environment is ubiquitous with objects, materials, and humans, which all possess rhetoricity—the quality of being rhetorical, rhetoric's animation, or possessing the capacity to communicate, perhaps persuade, or simply leave open to human interpretation and the very human inclination to connect the dots they can see and feel, given any other unknown or invisible environmental, material, or affective factors. I remain committed to my claim in the discipline of writing studies to attend to the affective, rhetorical spaces that exist between humans and nonhumans. Giving language to the seemingly neutral, invisible spaces between bodies, human and nonhuman, can offer insight into what matters to writers regarding their relationships with the materials they write on, with, around, about, and because/in spite of. I must acknowledge the dissonance these strands of thought and research create in/for scholarly discussions and debates surrounding materiality, rhetoric, and competing epistemologies, as I work within and against assertions that conflict with my own.

LANGUAGE AND POWER

Ultimately, language is wholly inadequate to respond to an affective, undefinable range of interactions between materials, both human and nonhuman. This complicates the teaching of writing in that without the acumen of language to account for and reveal the affective relations within each (often invisible) material interaction, writing could be condemned as nothing more than an artificial way of viewing the world, of constructing a reality that is built for those who use it for their own means and, ultimately, ends (Katz 1992). In essence, language in its seemingly infinite uses, is limited and quite finite because it cannot account for that which no language is sufficient. Words matter deeply, and language is a convenient human ability that embodies the tension between rhetoric and reality. Words are both instructive and demanding; from the lofty position of the paper to digital iterations in the form of websites, social media platforms, and government documents. They are dually responsible for creating coherence and imparting division among people and ideas. Words cannot be controlled after they are set in pencil or ink or through digital publication. They set themselves into motion with both intended and unintended consequences. Words possess the capability to produce enormous depth and reach—they shape the way we think, act, and engage in the world and with others. The human reliance on language maintains our stance that we are superior to all other forms of matter, human and nonhuman. Language reinforces human power because words *do* act. The language we adhere to most devotedly is that which is most comforting, that which is familiar, and that which agrees with our perception(s) and realities, dreams, and fundamental feelings about who we are or who we aspire to become. We must recognize that while we may understand our humanity as being something defined by the ability to use language, we actually must understand who we have the potential—the capacity—to become if we consider material, nonlinguistic agents that are not immediately evident yet are ever-present.

References

Alaimo, Stacy, and Hekman, Susan J. 2008. *Material Feminisms*. Bloomington: Indiana University Press.

Aristotle, and George A. Kennedy. 1991. *On Rhetoric: A Theory of Civic Discourse*. New York: Oxford University Press.

Barad, Karen. 2003. "Posthumanist Performativity: Toward an Understanding of How Matter Comes to Matter." *Signs* 28 (3): 801.

Barnett, Scot, and Casey A. Boyle. 2017. "Introduction: Rhetorical Ontology, or, How to Do Things with Things." In *Rhetoric, through Everyday Things*, edited by Scot Barnett and Casey Boyle. Tuscaloosa: University of Alabama Press.

Brummett, Barry. 1979. "Three Meanings of Epistemic Rhetoric." SCA Convention, November.

Clary-Lemon, Jennifer. 2019. "Gifts, Ancestors, Relations: Notes toward an Indigenous New Materialism." *Enculturation*. https://enculturation.net/gifts_ancestors_and_relations.

Coole, Diana H., and Samantha Frost. 2013. *New Materialisms: Ontology, Agency, and Politics*. Durham, NC: Duke University Press.

Hawhee, Debra. 2017. *Rhetoric in Tooth and Claw: Animals, Language, Sensation*. Chicago: University of Chicago Press.

Ingraham, Chris. 2018. "Energy: Rhetoric's Vitality." *Rhetoric Society Quarterly* 48 (3): 260–268. http://doi.org/10.1080/02773945.2018.1454188.

Katz, Steven B. 1992. "Ethic of Expediency: Classical Rhetoric Technology, and the Holocaust." *College English* 54 (3): 255–275.

Kennedy, George A. 1992. "A Hoot in the Dark: The Evolution of General Rhetoric." *Philosophy and Rhetoric* 25 (1): 1–21.

Martin, Emily. 1991. "The Egg and the Sperm: How Science Has Constructed a Romance Based on Stereotypical Male Female Roles." *Signs* 16 (3): 485–501.

Rickert, Thomas J. 2013. *Ambient Rhetoric: The Attunements of Rhetorical Being*. Pittsburgh, PA: University of Pittsburgh Press.

Said, Edward W. 1978. *Orientalism*. New York: Pantheon Books.

Scott, Robert L. 1967. "On Viewing Rhetoric as Epistemic." *Central States Speech Journal* 18 (1): 9–17. https://doi.org/10.1080/10510976709362856.

Vitanza, Victor J. 2003. "Abandoned to Writing: Notes Toward Several Provocations." *Enculturation* 5 (1). http://enculturation.gmu.edu/5_1/vitanza.html.

von Petersdorff, Anne. 2018. "Wanderlust, Cuerpos en Transito." Presentation of hybrid dissertation: The University of Texas at El Paso. https://www.annevonpetersdorff.com/hybrid-dissertation/.

Worsham, Lynn. 2015. "Moving beyond the Logic of Sacrifice: Animal Studies, Trauma Studies, and the Path to Posthumanism" In *Writing Posthumanism, Posthuman Writing*, edited by Sidney I. Dobrin. Anderson, SC: Parlor Press.

Zabrowski, Katie. 2016. "Alinea Phenomenology: Cookery as Flat Ontography." In *Rhetoric, through Everyday Things*, edited by Scot Barnett and Casey Boyle. Tuscaloosa: University of Alabama Press.

11
Writing into Inclusion from the Margins

ANTONIO BYRD

Introduction

I didn't intend to become a writing studies scholar; I wanted to write fiction. I had written stories since the third grade, but I took a career in fiction writing seriously in high school when Hollywood began releasing film adaptations of the most well-known fantasy writers of the twentieth century: J. R. R. Tolkien and C. S. Lewis. In summer 2002, George Lucas returned with his second film in the *Star Wars* prequel trilogy *Star Wars: Episode II–Attack of the Clones*. I consumed the books and films with relish and transformed my excitement for the pop cultural moment into three sci-fi/fantasy novels of my own (still saved on floppy disks and thumb drives!). I even submitted them for copyright registration! During orientation for first-year students in college I proudly claimed myself to be an English major, and that's when my mother stopped me: "You can't major in English," she said. "There's no guarantee you'll make a living off writing, so you should teach. Get something that will put food on the table while you write on the side." The practical and safe way to high-quality life, especially financial independence, was in teaching writing, not being a writer.

My curriculum for secondary education with an emphasis in English included multiple sections of English and American literature, and special

https://doi.org/10.7330/9781646426331.c011

topics in literary study took deeper dives into similar cultures. Meanwhile, I wrote short stories, shared them at open mic sessions hosted by the English Club, and even published a few pieces in my alma mater's literary journal. My mom was right: maybe I could do both of these things! But teaching descriptive writing to seventh-graders during my class observations in a magnet school led me to writing studies. After showing my students different photos to explain multiple ways a writer does descriptive writing, I asked them to write their own paragraph and then workshop their work. The controlled chaos felt more like home than the world of literary analysis and standardized tests.

I look back on my initial career trajectories as my believing in the innocence of literacy, that language was a matter of wrestling meaning to the page. If the meaning is clear and delights readers, I too could be a famous writer with film adaptations. If my students write with clear meaning, they too could be strong writers. However, that understanding of language's neutrality—that it does work autonomously (Street 1984)—only confirmed how whiteness hides the impact of race and language on marginalized communities. Guided by pop culture and college study, I was led to "reproduce the academic manners of the elite" (Fox 2009, 127), otherwise called the "White racial habitus . . . sets of durable, flexible, and often invisible (or naturalized) dispositions to language that are informed by a haunting Whiteness" (Inoue 2017, 27). In other words, the innocence of literacy led me to reinscribe Eurocentric fantasy stories.

My interaction with coursework in doctoral studies that highlight white scholarship felt familiar, then, but only because I continued what I always knew: accept the call to "perform their academic identities through standard conventions, disciplinary legitimacy, and other actions that ensure professional access, . . . a colonizing practice that results from the historical colonization of intellectual space in which BIPOC [Black Indigenous People of Color] often find themselves" (2017, 40). Ultimately, BIPOC scholars and teachers are called to, in the words of other authors in this collection, "manufacture some sort of whiteness" (Tellez-Trujillo, chapter 7), which requires the "obedience, silence, and diligence" that Cynthia Johnson, also in this collection, finds necessary to make it in writing studies (chapter 6). Doing so rewards BIPOC scholars from white institutions, creating the illusion of doing good work, when it really misaligns with calls for social justice, inclusion, and accountability across multiple social identities.

My autoethnographic account continues where I leave off at the conclusion of my undergraduate studies. I uncover how witnessing racial injustice helped create a pathway to the margins and to understand literacy's role in antiracism. I describe three key moments that led to my staking a claim in writing studies from a Black perspective. First, I used my white racial habitus of writing for a lesson plan that assumed students would be emotionally disoriented with digital multimodal composition. Second, the death of Mike Brown uncovered the limitations of my initial framework for writing and that the pathway to writing studies was claiming whiteness itself, leaving racial injustice unchallenged. The final moment describes going to the margins of writing studies as a rhetorical tactic for claiming writing studies; that is, learning the Black freedom struggles—"struggles engaged in by subjects racialized as black to mark their humanity, make legible their legal and extralegal exclusion from societies built by their labor, and form new worlds by transforming and creating inclusive and equitable social conditions" (Johnson 2018, 58)—documented in past scholarship. Grounding myself in the margins creates inclusive teaching and research that responds to racial injustice.

Moment #1: Teaching a White Racial Habitus for What Counts as Writing

When I finished my bachelor's degree in education in December 2011, I had become more interested in writing pedagogy than teaching middle school and high school literature. A return to my alma mater would, I had hoped, give me clarity on how to teach writing well in secondary education. I had not thought about teaching writing in higher education at that point. The master's program I attended, similar to my undergraduate coursework, still centered white literature, leaving me to theorize writing pedagogy on my own as a graduate teaching instructor for first-year writing and the writing center. However, an additional avenue came to my attention when a friend introduced me to the director of the composition program who had accepted him for an independent study. For the first time ever, I heard the words "composition and rhetoric" and discovered there was a formal scholarship on teaching writing in the classroom. The director invited me to participate in an independent study as well. To be in a community of other graduate students, she also invited the two of us to visit her creative nonfiction course each week.

The director encouraged me to use my interests and curiosities as a guide into composition and rhetoric. I thought about my leisure use of the Internet: spending hours on 1up.com, a now-defunct social media and news site for video gamers, and blogging about philosophy, Christian theology, and Japanese animation on Wordpress. I realized that the proliferation of digital technologies and the Internet was changing writing, yet instructors still taught writing as print-only. I found compelling Kathleen Yancey Blake's call to study the digital composing practices of students outside the classroom to develop "new models of composing" and "a new curriculum supporting those models" (Yancey 2009, 8). With my discovering the need for new directions in writing instruction and my background in secondary education, I was ostensibly on firm footing to apply the knowledge of writing studies in the classroom setting.

To explore how first-year writing could evolve into digital multimodal composition, I developed a ten-day lesson plan on remediating creative nonfiction essays into digital stories. I designed the unit to take place later in a semester after students had discussed, written, and workshopped creative nonfiction essays. Scholars before me had already suggested that text remediation offered students new ways to think rhetorically about communication in new media formats and better prepare them to think more deeply about the construction of their essays (Palmeri 2012). Text remediation, I thought, would help mitigate the discomfort that writing instructors may feel when teaching and assessing digital multimodal assignments (Borgmann 2019). Cheryl Ball's suggested revisions of the WPA Outcomes for First-Year Writing to include multimodal composition instruction (Ball and Moeller 2007) further helped me justify remediation to writing program administrators concerned about the place this work had even in a first-year writing course. Although I had been teaching first-year writing for a year by then, this project was my first opportunity to frame writing on my own terms.

My project focused on technical skills and imagined students as emotionally unprepared for this new way of writing in a classroom setting. While I trusted them to write creative nonfiction essays on their own, I framed text remediation for instructors as an emotional trauma. For example, well into the unit, lesson plans directed instructors to ask students, "Was [the text remediation process] *hard?* . . . *Or maybe the whole process was a breeze.*" In another lesson later in the unit, I probe for levels of difficulty and unease: "*Is it easy? Is it difficult? Do they feel comfortable with what they are doing so far?*" I write in the lesson, "The instructor should encourage [their] class to keep moving forward, whether the students are *frustrated or cruising* through it all." To further mitigate my imagined students' disorientation, in the first two classes of the unit plan, I reduce writing for remediation as a mere process of matching pieces of one genre with the next. Students produce storyboards and a script to select key elements of the essay that can reappear as a collection of the other available means of communication—video, sound, gesture, and images.

Given that students are often more prepared to use alphabetic literacy, I write that the instructor can "remind [the student] that if they can tell a story on paper, there's no need to make a digital story. In other words, students must take advantage of what a digital story can do. Sometimes the images, sounds, and animation can tell the story better than words can." Thus, writing is compartmentalized and simplified to help students make these otherwise difficult rhetorical moves in text remediation. This focus on navigating technology culminated in other activities that further emphasized learning the technical skills of digital multimodal composing: an introduction to the variety of software students can use to create their digital story (i.e., iMovie, Final Cut Express, Garage Band, Audacity, Windows MovieMaker, etc.) and an entire class meeting in the computer lab where students construct their digital stories under the instructor's supervision.

My final project appeared to have effectively synthesized my previous background in education with my current understanding in writing as multimodal to develop what I recognized as a pedagogically innovative approach to composition in the twenty-first century. Thus, I thought I was achieving my goal for returning to graduate school: to learn how to teach writing effectively. The independent study was a moment of pride and a moment of crossing a threshold into writing studies. However, in retrospect I observe the consequences of my framing literacy as innocent, unproblematic, objective; my project understood digital text remediation as "simply a matter of imparting technical skills rather than facilitating development of diverse and significant literacies" (Sheppard 2009, 123). Wrestling with tech as a pathway to finding clear meaning and delight with audiences from text to remediated text left de-emphasized the cultural context of writing; my empathic framework for teaching multimodal composition didn't take into account students as capable, curious, and rhetorically savvy writers, nor did it encourage students to see themselves in this way. I also focused less on how diverse lived experiences and rhetorical awareness informed the production of multimodal texts in communities. I had misread Blake's call to study students' digital composing outside the classroom context, and thus even excluding my own digital composing on social media. As I began doctoral study, I had a "felt difficulty" (Takayoshi, Tomlinson, and Castillo 2012, 100) with the white racial habitus of writing I had adopted and then extended into my advice for writing instructors. The severe limitations of this framework for addressing racism became apparent in the work of state-sanctioned racial violence.

Moment #2: Mike Brown, Ferguson, and Madison, WI: The Bridge to Racialized Literacies

Accepting the invitation to attend the University of Madison-Wisconsin for my PhD in composition and rhetoric was a win for Black representation. As an undergraduate I first heard about the importance of representation in doctoral programs from an education professor; she explained to me that if I went for my PhD, the school would pay me for my education, especially as a Black student. As an undergraduate I worked as a student assistant in the Office of Diversity and Inclusion. The chancellor of Diversity and Inclusion insisted that more Black men should get doctoral degrees; the numbers are poor and the representation matters, he reasoned, and insisted I consider a PhD program. Finally, after I decided to continue my education at UW-Madison, a friend and I discussed the role of the Black public intellectual, which I was presumably going to become. Is there a responsibility to represent and speak for the Black community, like Michael Eric Dyson or Cornel West? Black representation made sense to me. I believed in the importance of having a seat at the table, and I felt pride in my achievement. A PhD seemed so distant and impossible when I was in my first year of college and my mother had asked me, "Are you going to get your Master's?" as I struggled to pass pre-algebra. But, as Andre Brock once said during a webinar called Anti-Blackness and Technology in 2020, representation will not save us. In Moment #2, I reflect on how protests in Ferguson, Missouri, against the police shooting and murder of Mike Brown shook apart Black representation; this event was also a counterstory (Martinez 2020) to my framework of writing as skills-based and ultimately affective and my relationship with writing studies as a Black person.

Darren Wilson shot and killed eighteen-year-old Mike Brown on August 9, 2014. That was two days before Brown would have attended his first semester of technical school. By the time I had arrived in Madison, Wisconsin, in mid-August, protests on the streets of Ferguson and the militarized response to them were already unfolding. In a television interview, Mike Brown's mother Lezley McSpadden told a reporter, "You know how hard it was for me to get him to stay in school and graduate? You know how many Black men graduate? Not many!" (Chang 2016, 89). Theresa Perry, Claude Steele, and A. G. Hillard III argue that the African American literary tradition includes an "indigenous African-American philosophy of education" that came "out of [their] collective experience with learning and education, and all that that implied, . . . that was passed on in oral and written narratives" (2003, 12). This literate and rhetorical practice has been one reason Black communities persist despite racism. But Mike Brown's mother also identified how the philosophy of education for Black people is challenged not by other philosophical schools of thought but by white supremacy armed with police brutality. We've seen this before Ferguson, when James Crowley arrested Henry Louis Gates Jr. for breaking into his own home, and thereafter when Stewart Ferrin arrested Ersula Ore (a moment Ore would later discuss during a visit to UW-Madison during my time in graduate school). Two Black academics served injustice.

Thus, the narrative of Ferguson disrupting notions of Black representation was nothing new. However, because it happened as I embarked on my first semester of doctoral study, it was formative for my own thinking. My claim in writing studies from the perspective of neutrality, technical knowledge, and getting comfortable with digital multimodal composition offered little protection for me; it was better suited for white racial habitus, the race that never needs to name itself because it is the default and the standard. The justice system would notice my skin. But my literacy, and the scholarship I read to prepare for the fall, did not account for the skin and the social consequences associated with it.

What further underscored this point about literacy and Black representation failing to overcome racism was that I had moved to a majority-white progressive city that was economically vital and highly educated yet struggled with racial inequity. During my visit to UW-Madison before accepting my admission, graduate teaching assistants explained that while the white students had ambition, they also lacked "cultural awareness." My host supported this observation with his own stories, such as the time a white undergraduate roommate called him "nigga" while they play fought or how he never saw a Black person on campus unless they cleaned or served students food. My host later became my roommate when I moved to Madison. In the first few weeks of my arrival, we saw a white neighbor climb the stairs instead of taking the elevator with us to the same floor we all lived on. A few months after moving to Madison, I would find myself walking in a nice neighborhood to visit a friend and feel that if anyone saw me, a Black man, perhaps my bookbag would signal that I'm a student and thus am no threat to the white families living in those homes. But this everyday lived experience was in the backdrop of gross racial disparities in Dane County (Wisconsin Council on Children and Families 2016) and studies that found Wisconsin had the highest rate of Black incarceration in the entire country (Quinn and Pawasarat 2014).

Entering doctoral study was notable, a win for Black representation. Yet the Ferguson protests and my slow realization of Wisconsin's institutional racial inequality showed I had actually entered enemy-occupied territory. The irony is that only by leaving the South for a progressive community did I feel most aware of race and racism. I do not mean to say that there were no allies, because I had the blessing to build relationships with my peers, mentors, faculty, and community members. I do mean that white liberal progressive politics has its limitations, as Derrick Bell and other foundational critical race theorists noted decades ago (Bell 1992).

The shooting of Mike Brown, the Ferguson protests, and my own academic and social environment prompted new questions on how to continue the Black philosophy of education and achievement through writing and rhetoric amid racism. The definitive answer wasn't necessarily in my studies or in my prelims list, no clear response through the academic literacy I had developed. The framework for writing from my independent study, the one that focused on technical skills and strategies for addressing discomfort with new writing technologies and writing tools and brought me into writing students, reflected a Western-centric framework for research. This framework identified by Patricia Hill Collins (1989) applies to writing as well (Shelton 2020): ignore personal values and ethics, distance oneself from the object (in this case digital media software, as if a neutral tool) and, what most germane to my beginning conception of writing, detach oneself from emotions. I neutralize possibilities of understanding both digital multimodal composition and using the tools for composition for its cultural values, practices, and its consequences. Helping students detach from digital technologies to then conquer those technologies reflects a colonialist understanding of writing; one must control and make neutral the work. They were not meant to be used for addressing the macro- and micro-level forms of racism described above. I wondered what other frameworks of writing "seek justice and restore (and, in some cases, create) peace that reaches beyond the classroom walls"? (Winn 2013, 127). How might I use literacy to create "peacemaking circles" that "engage in dialogue; [and] demands collaboration and consensus" and "[eliminates] hierarchies based on academic prowess or social and cultural capital" (Winn 2013, 128)?

The answer was not clear to me in summer 2014, and I could not articulate what I was witnessing on television and encountering in my everyday life as well as I can now in this account. As I sat in my first seminar, I thought about my faculty mentor's advice: your stated research interest in your personal statement is not a contract; the personal statement shows potential and ways you and faculty may connect. Instead of holding to multimodal composition, explore. Be open. In light of this advice, I pondered about the contributions of Black people to writing studies. With this question, I could look back on my literacy history described (the innocence of literacy), which was absent of intellectual ancestors who theorized not only race but also Blackness in the context of white supremacy and its evolution over the centuries. The simple thought of what African Americans have contributed to composition and rhetoric put me on a path toward studying the noticeable "stream" (Royster 2000, 5) and known oceans of Black freedom struggles for literacy.

A deep dive into understanding how and why literacy mattered to racial difference could be a starting point into a lifelong process of unlearning literate whiteness and relearning how the margins spoke to and about writing studies. A return to school for a PhD, then, was not a *continuation* of learning; my return to school marked the beginning of learning anew what the marginalized said and did with writing. I began a journey from the defaulted center to the little-acknowledged margins to find a new approach to claiming the field.

Moment #3: Documenting Black Freedom Struggles in First-Year Doctoral Study

In her article "Markup Bodies: Black [Life] Studies and Slavery [Death] Studies at the Digital Crossroads," Jessica Marie Johnson (2018) argues that Black freedom struggles "challenge reproduction of black death and commodification, countering the presumed neutrality of the digital." Although Johnson analyzes Black digital practices of curation, the statement overlaps with Black freedom struggle for language and literacy. My first year as a graduate student was an exercise in finding these struggles at the margins of writing studies to strengthen my sense of communal justicing and critical language awareness (Gere et al. 2021). I began learning how Black/African American scholars contributed to writing studies. In understanding this contribution, I grounded myself in efforts to legitimize Black linguistic practice across generations. Then I recounted a project in spring 2015 that shows how Black beliefs in the power of literacy could inform current efforts to democratize opportunities to learn computer programming, possibly preventing the repeat of racial stratification and opportunity hoarding in public schools. Learning about the ways Black people took up literacy for freedom would help me revise myself as a Black scholar and a teacher of writing and claim a space in writing studies crafted for and by me.

A GENEALOGY OF RACIST DISCOURSE ON
AFRICAN AMERICAN LANGUAGES

My first seminar in fall 2014 was conceptually based on Malcolm Gladwell's "tipping points" (Gladwell 2000) to track paradigm shifts in histories of writing studies. However, these paradigm shifts are not always identified as such until much later when we witness its "downstream" impact. An example from this course was the 1966 Dartmouth Seminar as a major tipping point in writing studies; its gathering of scholars helped move composition from teaching writing as five-paragraph theme essays to studying the "individual learning and processes of mind" (Nystrand 2005, 13). Although a significant meeting, then, we understand its value until after evaluating the robust scholarship produced in the decades since. To apply tipping points and their downstream impact, the first seminar paper assignment asked that I argue why a historical event, or a series of historical events, was a tipping point in the field. This assignment allowed us to strike a new and/or alternative historical account of the field that would necessitate a new interpretation of our goals as scholars and teachers (Ruiz 2016).

On the heels of Mike Brown's death and the Ferguson protests, I pursued my initial question: How have Black people contributed to the field? My way into Black freedom struggles began with Keith Gilyard, who quite literally responded to my question by restating it in his article's title: "African American Contributions to Composition Studies." Gilyard (1999) "trace[s] a line of thought from early rhetors and scholars to contemporary researchers, thinkers, and practitioners that both emphasizes *critical pedagogy and values Black culture, especially its vernacular language*" (626, emphasis mine). This tracing of thought, Gilyard continues, is necessary to "produce critical and astute African American students" (626). Those African American students included myself. From Fredrick Douglass to Martin Luther King Jr., Gilyard offered me a sufficient introduction on how Black people formed a legacy of liberation through the uses of rhetoric and literacy, not unlike Perry's Black philosophy of education. Yet it was his centering white resistance to Black dialect and deficit thinking of Black students that resonated with other scholarship I was reading for a separate seminar that introduced me to the foundations of the field. The 1979 Ann Arbor court case, in which the Supreme Court ruled that public school teachers must accommodate Black linguistic practices, overlapped with my reading of Vershawn Ashanti Young's (Young 2009, 2014) advocacy for code-meshing and Paul Matsuda's (2014) call for more careful attention to translingual writing in composition. I saw a thread stretching from the past to the present, from deficit thinking about language, and Black language in particular, to translingualism.

My seminar paper "A Genealogy of Discourses on the Ebonics Debate" synthesized these conversations I encountered in my courses with additional research to apply Bakhtin's authoritative discourse. Authoritative discourse is the official narrative and knowledge told from the dominant power structure; this discourse shapes our language and how we interact with our social reality (Bakhtin 1981). "A Genealogy" discusses how other discourses challenge official perspectives on Black linguistic practices; tracing the genealogical roots of this discourse culminates in the 1979 Ann Arbor court decision as a tipping point in histories of linguistic racism.

My study began in the late nineteenth century when Paul Laurence Dunbar's poetry collection *Majors and Minors* (1895) was published to a rave review by William Dean Howells. Dunbar's poetry was written in both "standard English" and Black dialect, and it was the poetry of Black dialect that made Howell believe he was "in the presence of a man with a direct and fresh authority to do the kind of thing he is doing" (n.p.). Howell's praise represented white fascination with Black dialect and not a recognition of Black dialect's rhetorical power. That perspective, I wrote, would be a longer battle spanning from Lorenzo Dow Turner's study of the Gullah dialect in 1929 to William Labov's (1972) conclusion that speakers of Black English are better "narrators, reasoners, and debaters than middle-class speakers who temporize, qualify, and lose their arguments in a mess of irrelevant details" (213–214). These conversations on linguistic racism remained siloed from teaching until it spilled into public discourse. I selected the Ann Arbor case as a public institution that, backed by the expertise of linguists and later the "Students' Right to Their Own Language," spoke to the truth of unconscious bias in teachers when they first encountered Black students and their language practices. Investment in translingual writing offers opportunities to correct the field's own linguistic racism and develop language to publicly engage conversations on what kinds of Englishes matter in the workplace and beyond.

The genealogical discourse captured above highlights how the language of white racial habitus goes unquestioned—the true "neutral," while Black linguistic practice is political and social, a mystery that must be sorted out despite its derivation from the English Black slaves encountered in the South. The Ann Arbor case was one of many assertions for antiracism in public schools. White teachers, the research shows, were encouraged to either assimilate Black students into white society or ignore their presence in the classroom, thus institutionalizing a *de facto* segregation among students.[1] Both racist policies for education reflected the material consequences of language and literacy and effort to imbalance the relationship between white racial habitus and Black linguistic justice. Meanwhile, the Ann Arbor case advocates for a change in these policies for teacher education and pedagogy. It also highlights, I think, the downstream impact these histories can have on writing studies. We can evoke harmful past ideologies in language in our present without knowing where they came from. But more troubling is how ideologies on the "problem" of Black and Brown linguistic practices carry forward more than antiracist practices in writing studies can. Despite a variety of calls for linguistic justice from "Students' Right to Their Own Language" to debates on code-switching and code-meshing to translingualism, the ideology of white racial habitus remains. What we internalize and welcome in our house becomes the dish we serve to students as "nutrition": a field that feeds harmful internalized anti-Black linguistic racism (Baker-Bell 2020).

The range of sources I pulled from in this seminar paper suggest that pathways into writing studies involve an interdisciplinary coalition among education scholars, critical race theorists, linguists, and writing studies activists to bring about critical language awareness (Gere et al. 2021) for ourselves, our students, and our colleagues among students and teachers. This comes by way of intentionally studying the historically excluded writing and rhetorical practices of rhetoricians and writers in response to race and racism. That history provides a blueprint for protecting the possibilities for Black achievement in writing studies, and the achievements for other BIPOC. However, ongoing white languaging underscores tensions between the interests of stakeholders who valorize white racial habitus (university administrators, alumni, employers, curator's boards, etc.) and the intentions and mission of writing studies potential to make the study and teaching of writing for equity. As much as I would like to burn everything to the ground, I know doing so may put myself and BIPOC students in worse conditions than what's already offered in higher education. To that end, I collaborate with others' antiracist research agendas, caucuses, presentations, and community engagement efforts.

THE LINE OF COLOR IN COMPUTER PROGRAMMING

In fall 2014 I understood the necessity to critique writing and rhetoric for its cultural and political violence against Black linguistic practice. The following spring semester, I shifted to digital literacies and race for a seminar on Black intellectual thought and education. This course covered the perspectives of well-known Black scholars—W. E. B. Du Bois, Marcus Garvey, Anna Julia Cooper, among others—that can and do inform teaching in public schools. I had come full circle in my studies as a teacher, from coursework that taught practical instructional strategies and classroom management to exploring the social and cultural ideologies that operated behind those strategies, some of them more detrimental to BIPOC students. The seminar project had only two requirements: it must be twenty pages or more and "of scholarship," or, as my professor said, related to curriculum and instruction.

Still interested in digital multimodal composition, I explored the relationship between race and technology. Adam J. Banks (2005) introduced me to an ongoing concern: "How is it that writing studies and technical and professional communication disciplines demonstrate social and cultural awareness of meaning-making yet overlook race and technology? How is it that the ways of African Americans and other people of color use digital technologies are entirely neglected even in the few conversations that do take place?" (15). His question on race and technology overlapped with Annette Vee's call to interrogate the social implications for requiring computer programming in public schools. By drawing parallels between histories of alphabetic literacy and coding literacy, Vee found that calling coding a literacy was not just a rhetorical argument: coding *is* a type of writing. She notes that "computer programming is often portrayed as asocial, or purely technological" when they are actually "social technologies in their design and deployment" (Vee 2013, 55). Finding calls to include the lived experiences of racially marginalized people in the design of technologies and coding and having understood in the previous semester that writing and literacy was about power and inequality, I sought to map Black intellectual thought on education onto computer programming. My interests in digital multimodal composition began to center the cultural practices of literacy and undo years of literacy as white racial habitus.

In my seminar paper "Write the Color-Line in Coding as Literacy," I found that theories of Black intellectuals on education and literacy from the nineteenth and twentieth centuries could inform perspectives on the artistic and rhetorical functions of computer programming. Black people inherited a tradition from slavery and Jim Crow that espoused literacy as tools for liberation. Because education in the United States is already unequal, the introduction of computer programming throughout public schools would most likely continue those policies of stratification. A seminar paper that first names Black philosophies of education and then suggests how the expansion of coding could adopt these philosophies in their curricula and teaching would "help us learn from the past, to recognize a repeated mistake in progress, to stop the current trajectory, and propose a new path forward" (Byrd 2015, 8). Maria Bibbs's dissertation on Black people's pursuit of literacy during the Progressive Era taught me a distinction between the Black literacy myth and the white literacy myth. While Harvey Graff (1991) quantifiably questions literacy producing real advantage in nineteenth century North America, Black people believed in its ability to liberate and, in the lives of many ex-slaves, it followed through on that promise. The histories of Black excellence that Theresa Perry articulated comes full force in W. E. B. Du Bois's call for the United States government to "step in and wipe out illiteracy in the South" (375). Meanwhile, the Harlem Renaissance[2] staked a claim to producing Black aesthetics and culture to undo centuries of false narratives about Black people through the white gaze and compete with European aesthetic and artistic traditions and counter racist stereotypes of Black people while refusing to assimilate into mainstream society. For this reason, Black artists and writers attempted to capture the full variety of Black experiences in the United States in ways white writers of that time could not. These notions signaled an effort to claim literacy as an especially unique Black property cultivated for their own community in the face of white supremacy.

These concerns, I argued, echoed claims that computer programming is a symbol system for communication and rhetoric; thus, coding can express ideologies on our social realities. I noted that several twenty-first-century (white) scholars and artists spoke of the multiple ways computer programming shared similar functions to reading and writing. If Black intellectuals saw literacy as tools for liberation and white cultural critique, critical code studies advocated for reading computer code to conduct similar cultural analysis. Mark Marino (2006) identified several scholars doing such work, such as Noah Wardip-Fruin, Nick Montfort, Michael Mateas, and Rita Rayley. Such critiques could remove notions that coding is neutral and show they are fused with biases that structure everyday lived experiences. Although critiquing racism remained missing from this scholarship, the scholarship implied that there was a possibility for antiracist coding.

In addition, I discovered that code could be used for cultural production and aesthetics. Other scholarship demonstrates that computer code contains meaning in two ways: in what the lines of code actually say and what kind of software those lines of code create. This suggests computer programming is a type of writing that produces digital literature and art. Weird language, obfuscate code, and even code poetry reinvents functions of coding away from serving capitalist technocratic dreams of profit and data exploitation. Instead, they identify with hacker cultures and creative expressions through computer programming. Studying and using coding for cultural critique and art, I noted, reflected the so-called alternative ways with words found in Black literacy and rhetorical traditions. Black people have used literacy and rhetorical education to engage white society, to critique its institutional racism and everyday microaggressions and even what it means to be Black within Black communities; but my own people's language practices work outside "standard English" to express these critiques and stake out a space for Black aesthetics.

To zoom out from my own positionality in this narrative and examine how what I discovered in this seminar paper extends to pathways into writing studies for BIPOC graduate students, I'm reminded of Iris Ruiz's (2016) *Reclaiming Composition for Chicano/as and Other Ethnic Minorities*. In her work, Ruiz explains how Black people and Mexican Americans taught composition in normal schools in order to preserve language practices and cultures; only until the Civil Rights Movement did mainstream universities and the field of writing studies begin to produce scholarship and critical pedagogy that critiqued their own white racial habitus. In other words, mainstream writing studies is often playing catch-up with intentionally excluded people. Her history shows that we have been talking about having a seat, perhaps being the only ones at the table, for a long time. The dominant narrative suggests deficit thinking of Black people until there's an interest convergence (Bell 1980) and white society then becomes an ally. As if plights of BIPOC were suddenly discovered, when in fact they had always been present at the mic; someone just needed to turn the volume up. Similarly, my synthesizing the Black intellectual thought on literacy with the work of critical code studies demonstrates that writing studies could have been (and still can be) antiracist from its conception. That process begins with listening and interacting with the Black people and other marginalized people intentionally excluded and marginalized for their practices and critiques. Then leading with using their knowledge for gains in social justice and human rights. The narrative of "discovering" new ways of writing after white racial habitus has marched through the field does not have to be repeated in our training ourselves and other graduate students.

For example, this second seminar paper offered an opportunity to make race the center of conversation in the coding for all movement, which was at that point and perhaps still is young in its efforts for democratization. When I shared my findings from this second seminar paper with friends and then later presented on this research for the 2015 Conference on College

Composition and Communication, audiences saw potential in what was missing about coding literacy and its democratization: the paper, although flawed, provided a theoretical basis for extending critiques of digital divide rhetoric beyond Internet access in low-income Black communities. The next step for my research was tapping into a community of Black coders and asking what do Black people bring to coding? If Black scholars understood how reading, writing, and orality worked for Black communities, in what ways does computer programming work toward similar ends of liberation? How do those experiences help us not only understand coding literacy as a racialized type of writing but also help us produce antiracist policies and practices in computer science education? Given the ongoing push to expand computer science education and its promotion as "the new literacy," I find urgency in bringing an update to Black digital literacy, not from Internet access or social media platforms but with the actual Black freedom struggle of designing technologies for themselves and other racially marginalized people. In doing so, I add perspective to conversations on equitable computing education through the lens of literacy studies.

Rhetorical Tactic: Writing as Cultural Practice

In this autoethnography, I've documented a process of shifting my framework for writing from white racial habitus to a framework that understands writing for its variety across cultural contexts. The genres I wrote in response to the contexts and assignments that structured my graduate education are a textual history that allow me to make a retroactive assessment of my becoming into writing studies. As constructed in my memory, my independent study, and first-year seminar papers, my journey involved multiple prior experiences in education and literacy and observing the cultural and political contexts outside of my personal experience that shaped how I would navigate the world as a Black person, scholar, and teacher. My pathway into writing studies came out of a desire to respond to race and racism that felt very real and present for the first time in my own life; I sought answers to what writing studies could do to address racism and found that answer by blending histories of Black language and literacy with the concerns of the present.

While taking stock of how Black culture, history, media, intellectual thought, and current events helps writing studies articulate the ways writing serves community-building, sustainability, and professional activism within the field (Perryman-Clark and Craig 2019; Ruiz and Baca 2017;

Royster 2003; Kynard 2013), I do not mean to suggest that BIPOC graduate students and scholars must also create a research agenda about race and racism. Instead, we may consider what in our academic lineage we take with us into our research agendas and how to productively shift our focus to create a well-rounded and inclusive sense of what writing does. To always be uncomfortable with our comfort and with our research and ways of teaching writing to ensure writing studies is ready to respond to injustice within and without higher education. This interrogation involves inclusive citation practices, yes, but also reflection of our histories and our present, and what they mean for training practitioners of antiracist practices in writing studies.

I demonstrate the necessity to create opportunities where graduates can recursively examine how the collection of their academic lineage and the trajectory of their professionalization does or does not help them develop frameworks that are inclusive of racially marginalized literacies. I suggest a similar rhetorical tactic to mentor-scholars guiding graduate students in writing studies, focusing on locating voices marginalized yet central to the work of the discipline. How do these voices challenge their own notions of writing? What opportunities open when we find what has been and can be inherited from historically excluded communities without appropriation? How does an account of the past help extend new ways of producing and teaching writing as a type of knowledge?

Notes

1. Literacy as a civil right would be set back twenty years later with the Oakland Resolution against Ebonics in 1997. Community members and some Black leaders criticized the Oakland, California, school board for suggesting teachers and students would learn Black English. Black students needed to learn "standard English." The criticism prompted the school board to remove references to teaching teachers and students Black English in public schools. This regression from the conclusions of the Ann Arbor case helped perpetuate anti-Black linguistic racism.
2. The Black Arts Movement would have been a better example of Black cultural production. Unlike many artists and writers of the Harlem Renaissance, the Black Arts Movement did not rely on white financial support.

References

Baker-Bell, April. 2020. *Linguistic Justice: Black Language, Literacy, Identity, and Pedagogy.* New York: Routledge.
Bakhtin, Mikhail M. 1981. *Dialogic Imagination: Four Essays.* Austin: University of Texas Press.

Ball, Cheryl, and Ryan Moeller. 2007. "Re-Inventing the Possibilities: Academic Writing and New Media Literacy." *The Fibreculture Journal*, no. 10. http://ten.fibreculture journal.org/wp-content/dynmed/ball_moeller/index.html.

Banks, Adam. J. 2005. *Race, Rhetoric, and Technology: Searching for Higher Ground*. Urbana, IL: National Council of Teachers of English.

Bell, Derrick. 1980. "Brown v. Board of Education and the Interest-Convergence Dilemma." *Harvard Law Review* 93 (3): 518–533.

Bell, Derrick. 1992. "Racial Realism." *Connecticut Law Review* 24 (2): 363–379.

Borgmann, Jessie C. 2019. "Dissipating Hesitation: Why Online Instructors Fear Multimodal Assignments and How to Overcome the Fear." In *Bridging the Multimodal Gap: From Theory to Practice*, edited by Santosha Khadak and J. C. Lee, 43–68. Logan: Utah State University Press.

Byrd, Antonio. 2015. "Write the Color-Line into Coding Literacy." Unpublished manuscript. Microsoft Word file.

Chang, Jeff. 2016. *We Gon' Be Alright: Notes on Race and Resegregation*. London: Picador.

Collins, Patricia Hill. 1989. "The Social Construction of Black Feminist Thought." *Signs* 14 (4): 745–773.

Dunbar, Paul Laurence. 1895. *Majors and Minors: Poems*. Toledo, OH: Hadley and Hadley.

Fox, Tom. 2009. "From Freedom to Manners: African American Literacy Instruction in the Nineteenth Century." In *The Norton Book of Composition Studies*, edited by Susan Miller, 119–128. New York: W. W. Norton.

Gere, Anne Ruggles, Anne Curzan, J. W. Hammond, Sarah Hughes, Ruth Li, Andrew Moos, Kendon Smith, Kathryn Van Zanen, Kelly L. Wheeler, and Crystal J. Zanders. 2021. "Communal Justicing: Writing Assessment, Disciplinary Infrastructure, and the Case for Critical Language Awareness." *College Composition and Communication* 72 (3): 384–412.

Gilyard, Keith. 1999. "African American Contributions to Composition Studies." *College Composition and Communication* 50 (4): 626–644.

Gladwell, Malcom. 2000. *The Tipping Point: How Little Things Can Make a Big Difference*. Boston, MA: Little, Brown.

Graff, Harvey. 1991. *The Literacy Myth: Cultural Integration and Social Structure in the Nineteenth Century*. New Brunswick, NJ: Transaction.

Inoue, Asao B. 2017. *Labor-Based Grading Contracts: Building Equity and Inclusion in the Compassionate Writing Classroom*. Louisville, CO: WAC Clearinghouse; University Press of Colorado. https://doi.org/10.37514/PER-B.2019.0216.0.

Johnson, Jessica Marie. 2018. "Markup Bodies: Black [Life] Studies and Slavery [Death] Studies at the Digital Crossroads." *Social Contexts* 36, no. 4 (137): 57–79.

Kynard, Carmen. 2013. *Vernacular Insurrections: Race, Black Protest, and the New Century in Composition-Literacies Studies*. New York: SUNY Press.

Labov, William. 1972. *Language in the Inner City: Studies in the Black English Vernacular*. Philadelphia: University of Pennsylvania Press.

Marino, Mark C. 2006. "Communities of Chatbots: A Survey of the Makers and Users of Conversation Agents." *New Media and Society*. Sage.

Martinez, Aja Y. 2020. *Counterstory: the Rhetoric and Writing of Critical Race Theory*. Champaign, IL: National Council of Teachers of English.

Matsuda, Paul K. 2014. "The Lure of Translingual Writing." *PMLA* 129 (3): 478–483.

Nystrand, Martin. 2005. "The Social and Historical Context for Writing Research." In *Handbook of Writing Research*, 1st ed., edited by C. MacArthur, S. Graham, and J. Fitzgerald, 11–27. New York: Gilbert.

Palmeri, Jason. 2012. *Remixing Composition: A History of Multimodal Writing Pedagogy*. Carbondale: Southern Illinois University Press.

Perry, Theresa, Claude Steele, and A. G. Hillard III. 2003. *Young, Black, and Gifted: Promoting High Achievement Among African American Students*. Boston: Beacon.

Perryman-Clark, Staci M., and Collin Lamont Craig. 2019. *Black Perspectives in Writing Program Administration: From the Margins to the Center*. Urbana-Champaign, IL: National Council of Teachers of English.

Quinn, Lois M., and John Pawasarat. 2014. "Statewide Imprisonment of Black Men in Wisconsin." *ETI Publications*, no. 7. https://dc.uwm.edu/eti_pubs/7.

Royster, Jacqueline Jones. 2000. *Traces of a Stream: Literacy and Social Change among African American Women*. Pittsburgh, PA: University of Pittsburgh Press.

Royster, Jacqueline Jones. 2003. "Disciplinary Landscaping, or Contemporary Challenges in the History of Rhetoric." *Philosophy and Rhetoric* 36 (2): 148–167.

Ruiz, Iris. 2016. *Reclaiming Composition for Chicano/as and Other Ethnic Minorities*. New York: Palgrave Macmillan.

Ruiz, Iris, and Damián Baca. 2017. "Decolonial Options and Writing Studies." *Composition Studies* 45 (2): 226–229.

Shelton, Celia. 2020. "Shifting Out of Neutral: Centering Difference, Bias, and Social Justice in a Business Writing Course." *Technical Communication Quarterly* 29 (1): 18–32.

Sheppard, Jennifer. 2009. "The Rhetorical Work of Multimedia Production Practices: It's More than Just Technical Skill." *Computers and Composition* 26 (2): 122–131.

Street, Brian V. 1984. *Literacy in Theory and Practice*. New York: Cambridge University Press.

Takayoshi, Pamela, Elizabeth Tomlison, and Jennifer Castillo. 2012. "The Construction of Research Problems and Methods." In *Practicing Research in Writing Studies: Reflexive and Ethically Responsible Research*, edited by Katrina M. Powell and Pamela Takayoshi, 97–121. New York: Hampton.

Vee, Annette. 2013. "Understanding Computer Programming as a Literacy." *Literacy in Composition Studies* 1 (2): 42–64.

Winn, Maisha. 2013. "Toward a Restorative English Education." *Research in the Teaching of English* 48 (1): 126–135.

Wisconsin Council on Children and Families. 2016. "Race to Equity: A Baseline Report on the State of Racial Disparities in Dane County." Race to Equity. https://racetoequity.net/wp-content/uploads/2016/11/WCCF-R2E-Report.pdf.

Yancey, Kathleen Blake. 2009. *Writing in the Twenty-First Century: A Report from the National Council of Teachers of English*. Urbana, IL: National Council of Teachers of English. https://www.ncte.org/library/NCTEFiles/Press/Yancey_final.pdf.

Young, Vershawn Ashanti. 2009. "'Nah, We Straight': An Argument Against Code Switching." *Journal of Advanced Composition* 29 (1–2): 49–76.
Young, Vershawn Ashanti. 2014. "Straight Black Queer: Obama, Code-Switching and the Gender Anxiety of African American Men." *PMLA* 129, no. 3 (May): 464–470.

SECTION IV

Conclusion

Embodying Stories

Writing Studies and Its Potential Paths

CHRISTINA V. CEDILLO

Upon reading this collection, the reader cannot help but be struck by the centrality of several crucial elements that we in writing studies (should) deem instrumental to our work. In every one of the rich and resonant essays found in this volume, we learn just how indispensable the act of storying is in relating lived experience and the valuable lessons to be drawn from the narratives we compose. In addition, we are reminded of the faculties of bodies as our most intimate of rhetorical apparatuses, and not just because bodies compose texts but because they transmit and receive information, take action, and engage in constellations[1] of meaning-making. For example, through the stories told here, we come to find that the term "blood, sweat, and tears" is not just a visceral metaphor but an actual corporeal aspect of building new programs and safeguarding the rights of those who work in them (Wood; Brooks, Dadas, Field, and Restaino; Ersheid, Konigsberg, McVeigh, Pearson, and Kahn). We also learn that histories are not written in words alone but are composed through human action "in conversation with" the material, the political, and the social (Bowen and Pinkert; Duffey; Zepeda). And we read about the all-too-common traumas imposed on members of marginalized communities and their effects on real people (Johnson; Tellez-Trujillo; Billingsley;

Rosas; Byrd). While every group, every individual, experiences discrimination in unique ways that we all cannot ever know, and every disciplinary program and its creators ascribe to very different histories, by reading these stories we are invited to bear witness to the effects of invisibilized labor practices and systemic injustice, and we are charged with acting in concrete ways that improve conditions for all.

As part of that work, I find this book's collective attunement to storying's biopolitical rhetorical power refreshing and necessary to claiming writing studies as a discipline. As a cultural rhetorician whose work focuses on how bodies are marginalized and how marginalization is reified in/through bodies, I value story as the primary technology that allows me to make sense of the world. Doing so entails recognizing that "the general term 'rhetorics' refers both to the study of meaning-making systems and to the practices that constitute those systems" (Powell et al. 2014), and that systems and their respective practices cohere through the stories that bodies tell. This research orientation is guided by an understanding that "culture and rhetoric [are] interdependent rather than stable categories" (Cobos et al. 2018, 141), imbricated in myriad ways that enjoin us to investigate systems, practices, bodies, and stories if we are to know as much as we can about any rhetorical situation. Engaging this vital nexus means getting messy, transcending the hygienic bounds of academic language that seeks to wrap things up nicely with a bow. Instead, we must make room for "texts, bodies, materials, ideas, or space knowing that these subjects are interconnected to the universe and belong to a cultural community [each] with its own intellectual tradition and history" (Riley-Mukavetz 2014, 109). Thus, I propose that we recognize story as a living, material, and bodily rhetoric that reorients our perspectives and praxes toward writing as *connection, emplacement,* and *enactment,* three types of rhetorical movement that can guide the future of writing studies in prioritizing equity and justice as central goals. These moves matter because recognizing that everyone and everything has a story means renouncing claims to any singular overarching narrative and understanding that stories compose and are composed by relations among parties (Cox et al. 2021), by where they "take place" (Powell 2012), and by what they do (Kerschbaum 2015).

To illustrate, and because I'm someone whose culture favors *plática*[2] (or being spoken with over being talked at), here I share some of my own journey into writing studies. I do so to highlight experiential and scholarly correspondences between my story and those of the authors featured here. In so doing, I aim to show the disciplinary potential of writing studies as a space of convergence where changemakers can work together to create spaces for engaging the world critically and ethically.

***

In the mid-aughts, after teaching English/language arts for a few years in California and then Texas, I decided to return to graduate school for my PhD in College Station. I had always dreamed about being a professor, but as the first person in my family with a graduate degree at all, the goal seemed lofty indeed. To tell the truth, I was terrified. I was applying to the same institution where I'd gotten my bachelor's degree and the place where I first experienced blatant racism and what I used to deem impostor syndrome but now recognize as the effects of white supremacy.

As an undergraduate, I attended a party exactly one time, with one of my suitemates and a few of her friends (all white women) who somehow didn't notice when another group of white women at the party whom they knew pointedly said hello to everyone in my group except me. Upon encountering those same white women in the bathroom, their "leader" looked me up and down, made a confused but pitying face, and then turned away, laughing with her friends. Still, I actually deliberated, was it was really racism? Growing up in a city that's now about 96 percent Latinx and was more so then, I recognized it when it looked like all the rich people in town somehow being white or white people receiving a promotion a parent had worked so hard for, but spatial regulation designed to teach you your place was something altogether new. Having spent my life in Latinocentric spaces, I was bound to learn, as Khadeidra Billingsley writes in chapter 8 of this volume, that "existing in a space that was not created or structured for your existence is exhausting and overwhelming." That night, I spent as much time as I could talking to the band hired to play the party, a group of guys from "liberal" Austin. Later, the owners of the house refused to pay and ran them off with threats of violence. As they left, one of the band members yelled out expletives at the party hosts who were now too drunk to notice. Seeing me, he waved and wished me a good night, calling me "the beautiful Mexican girl." (I went by "she" then.) I waved halfheartedly. I couldn't put my finger on why, then, but the supposed compliment rankled. My desire to be inconspicuous only increased. Why was my ethnicity relevant? Everyone was set on highlighting my Otherness, and people certainly did until I graduated and returned home.

My classroom experiences only exacerbated these matters, revealing that as much as the academy pretends to exist in some discrete utopic sphere where one is appreciated for their mind alone, identity is not something you leave at the door. Much as Karen R. Tellez-Trujillo explains in chapter 7, I found myself asking why nothing I wrote was good enough even when I tried my hardest to meet some unspoken expectations that everyone else seemed to grasp. Among my friends, I was *the* writer who people turned to for advice. Yet, where I had once been described as a brilliant student by my professors back home, now I was just mediocre. Like Raymond D. Rosas describes in chapter 9, I was "shook over norms I didn't even know existed, norms that were neither connected to the substance of the arguments I was putting down nor communicated explicitly." It didn't help that the depression that has characterized so much of my life was now manifesting in earnest, driving me to bed for days and leading my instructors to demand doctor's notes even though I didn't even know why I should be visiting a doctor. In one class, a professor asked for a note from my mother or else he would fail me for being absent four times in one semester, the semester that my only grandparent and the heart of our family died and that suicidal ideation became a common experience. The discomfort and shame kept me quiet, becoming yet another way for me to feel like a failure. Trying to pass as "normal" was painful, but that coupled with trying to fit into my whitestream surroundings and worrying endlessly about my future as someone from a working-class family was an excruciating amount of emotional labor that not everyone understands. (See Cynthia Johnson's chapter for an account of the different types of labor facing multiply-marginalized people.) And it's that emotional labor that the academy too often refuses to accommodate.

Yet here I was, ready to head to College Station ten years later though I swore never to return and once again risk several years of hell because the program I applied to had made me a generous offer. This time, however, my identity made me a catch since the university was taking steps to make their graduate programs more attractive to students from minoritized populations. I didn't realize then that throwing money at structural inequity does not equal inclusion, let alone progress or justice; I only knew they were paying. As several authors in this volume mention (Bowen and Pinkert; Duffey; Johnson; Billingsley), my career path was very much dictated as much by economics and personal circumstances as by choice, perhaps more so. My proof was in the award names themselves; those who were Black, Brown, Indigenous, or Asian received diversity fellowships—granted, the larger awards—while the merit fellowships went to white students. So off I went, knowing full well that I was being regarded as a token (see Billingsley), as proof of the altruism of

DEI (diversity, equity, and inclusion) programs, but knowing that if I didn't take this chance, I might not ever be able to afford PhD studies. Besides, I thought, perhaps things had changed.

But once again, I encountered several professors who discounted anything suggestive of "diverse" perspectives, like the one who loved to elucidate on whiteness as a medieval beauty standard and refused to admit that anyone in the Middle Ages might be queer. One only need read across the broad swath of those thousand years to realize that Europe was inhabited by more than just white people, that intimate relationships bloomed in close quarters like the monastery and convent, and that the privileging of homosocial relationships and religious androgyny meant that gender was sometimes regarded as more fluid than it is today (especially if, like me, you live in a state that is pushing anti-trans laws). However, we were dissuaded from writing on anything beyond textual language or from using anything but historicist lenses. Looking back, I find his reactionary approach bitterly amusing and sad. Ron Brooks, Caroline Dadas, Laura Field, and Jessica Restaino explain in chapter 4 that by welcoming queerness as a praxis that "embraces messiness, rejects categorization, and refuses binaries," novel paths of critical thinking could have opened up, bringing with them our ability to "[imagine] how things might be different," whether looking back at the past or gazing into the future in search of more inclusive perspectives. One thing that could have been different is my own relationship to queerness, which might have proven more secure as I wrote about people whose gender nonconformity felt familiar, but that professor's pedagogy precluded any opportunity to engage in challenging discussions or rethink our methodologies. Although Brooks et al. write about issues they confronted in building an independent writing program, I can see now how their struggles connect to mine as a student. Experiences like the one I had in that course are the ultimate outcome of institutional tendencies to frame writing as merely an expedient, service-oriented discipline rather than the locus of "exploration with positioning and positionalities" or "of how relationships are formed with those positioned differently from normative ideals" (Tellez-Trujillo, this volume). From my current vantage point, I recognize how those same forces permit, even reward, conventional approaches to reading and writing that leave many of us needing more. This disappointment was partly what spurred my switch from literature to rhetoric and composition.

Certainly, writing studies is not immune to fixed or conservative thinking regarding the future of the discipline. As Lauren Marshall Bowen and Laurie A. Pinkert show in chapter 1 through interviews with retired scholars and teachers who shaped the field's earliest trajectories, we find that people's affinity for one way of doing things, for certain methodologies and emphases over others, still affect how and what we perceive as the "pedagogical center" and what we view as "problematic." However, they also paint a portrait of a discipline that has "never *really* been stable," a space "forged by the fires of economic and political upheaval" that has always had to take people's real lives into account, whether willingly and deliberately or as a matter of pragmatic necessity. Due to fiscal and temporal exigencies, this need can result in some significant problems, but it can also lead to disciplinary introspection that informs vital calls for change. For example, Tara Wood notes in chapter 3 that our ability to engage in cross-disciplinary work has enabled the invention of stimulating, justice-driven learning, and yet we know that we are always constrained by "the state's articulation of competencies, content criteria, and the evaluative rubric they recommend for assessment purposes," all of which can hinder our efforts to ground our discipline in equity-oriented ideals. In chapter 2 Suellynn Duffey describes how universities create open admissions policies to make access to education more equitable and to increase minority representation. Yet, the same schools can establish systems and assessments steeped in racism and white supremacy, labeling students as "underprepared" and "remedial" rather than "differently prepared." The devaluation of equity work affects instructors, too, as attempts to improve conditions for contingent faculty continue to be viewed as outside of teaching, research, or service requirements. "Equity" itself proves a duplicitous term when deployed by institutions to further oppressive structure. Alison Ersheid, Lisa Konigsberg, Maureen McVeigh, Nancy Pearson, and Seth Kahn tell us in chapter 5 that teachers might even be coerced into changing their professional identities in order to secure job security and advancement, and even "labor-conscious faculty" can contribute to cultures that penalize vulnerable colleagues into disciplinary normativity. These conditions conspire to position writing studies right at the stressful intersection of revolutionary thought and rigid traditionalism.

But there are also the interstices, where, oscillating between what the past teaches us and what the future might hold, we can engage story as embodied knowledge in motion to cull crucial lessons. Whether imagined as a tornado (Duffey, this volume) that pulls matter together or some other erratic crucible that creates colloidal areas of concentration, the inherent instability of writing studies welcomes us to find areas of concord in the chaos. In surveying the disciplinary landscape, I'm struck by its resemblance to "rasquachismo," a Latinx term for the "combination of ingenuity, meaning making, boldness, and flair" that allows us to "recycle, upcycle, make do, and make new meaning through whatever available bits and pieces" while being more than a little subversive (Medina-López 2018, 3). I'm familiar with this scholarly cobbling because it explains why I did make the leap into rhetoric and composition and writing studies. My first ever PhD class was Histories of Rhetoric, and luckily my professor (now mentor and friend) was disability rhetorics scholar Stephanie Kerschbaum, who always asked us whenever we analyzed texts, "What can we do with this?" And those texts meant the books we read, yes, but also the cultural, political, material, and corporeal contexts we examined. In this class, we *were* invited to engage in "the examination of emotional and embodied responses and reactions to being marked as different, where students can think about the way that they have also marked others" (Tellez-Trujillo, this volume). This is the sort of learning that helps students apprehend how even written texts prove multimodal, finding purchase in living, breathing bodies, in material conditions, and constructions of temporality. This is the sort of learning that "changes how we and our students write with, around, about, in response to, and in the world" especially if we are to earnestly address the pressing concerns that endanger us and our earth (Zepeda, this volume). As Alison Wells Zepeda explains in chapter 10, we must recognize the objects and forces that shape identities while accounting for the experiences of vulnerable parties. Tracing these agential flows reveals to what degree forces like wealth, discrimination, and exploitation have influenced what and how we know (see Edwards 2020). Here is where I imagine that the disciplinary power of a socially engaged writing studies truly lies.

I stated earlier that we can think of writing studies' potential in terms of connection, emplacement, and enactment, but these possibilities require that we keep listening to stories as and within their full material and political contexts. It means getting real. As Rosas states in chapter 9, "Tropes of literacy are complex and problematic but, nevertheless, generative problem spaces for exploring how writing studies frames the promise of our work." Even in the midst of hostile conditions further exacerbated by alt-right rhetoric, police violence, and the COVID-19 pandemic, some scholars still argue that not everything we do has to be about social justice or contesting isms, addressing privilege and hardship, or colonialism and climate change. But as people who know better than most the importance of exigence and context, we should know that to ignore these elements of any rhetorical situation is to willingly erase vital connections that conjoin us to other human beings, to other denizens of the earth, to the land, air, and sky whose supposed separability from us has facilitated climate catastrophe. Stories, rhetoric, existence are emplaced, meaning they never take place in a vacuum; every new constellation brings with it a nexus of axiological hierarchies, positionalities, and power dynamics. To acknowledge this is not simply to "have an agenda," which we all do, but to recognize conditions that are and that affect everything from our identities to our writerly choices. What we then choose to do with that agency is up to each one of us.

For me, Antonio Byrd's work on the ways that writing studies can support community-building proves especially resonant. He states in chapter 11 that "we may consider what in our academic lineage we take with us into our research agendas and how to productively shift our focus to create a well-rounded and inclusive sense of what writing does," adding to that the need to "always be uncomfortable with our comfort with our research and ways of teaching writing to ensure writing studies is ready to respond to injustice within and without higher education." As a cultural rhetorician whose interest lies in the corporeal and experiential dimensions of knowledge, and as someone with my identity, I understand that said interest means "negotiation with and resistance between the discipline's histories of Whiteness and one's own racial context" (Byrd, this volume) as well as contending with the discipline's predilections for ableism, classism, queer erasure, and linguistic injustice. Because I choose to make room for myself in this discipline, I know that entails struggling against the historical tendencies to "[police] our borders as a field and [be] slow to embrace anything that wasn't just about first-year writing or about composition" (Pough 2011, 306). Because people like me are ever aware that "respectability will not save us in the streets or in relationships or in politics" and that "it will not save us in the academy either" (Banks 2015, 270), I admit I want—no, *need*—us all to perform those modest acts of recognition that bring change when enacted collectively. Ultimately, writing studies has the potential to reveal the unspoken, assist in rehumanization efforts by marginalized peoples, and make visible what and who has been invisibilized. However, that can only happen if we listen to the ancestral and future stories that beckon us onward.

Notes

1. Powell et al. (2014) define a constellation as a visual metaphor "that honor[s] all possible realities" and accounts for "different ways of seeing any single configuration within that constellation, based on positionality and culture," allowing all perspectives and relationships within the arrangement to matter.

2. Plática refers to conversation among relatives, either by blood or friendship, that sustains connections and "relational equilibrium" by providing a consubstantial discursive space where information can be shared and processed emotionally. It is "an expressive cultural form shaped by listening, inquiry, storytelling, and story making" that takes place both in the moment and over time as a series of conversations (Guajardo and Guajardo 2013, 160). Notably, it also relates to "tactile/cultural experience" (De La Trinidad et al. 2021, 38).

References

Banks, Adam. 2015. "2015 CCCC Chair's Address: Ain't No Walls behind the Sky, Baby! Funk, Flight, Freedom." *College Composition and Communication* 67 (2): 267–279.

Cobos, Casie, Gabriela Raquel Ríos, Donnie Johnson Sackey, Jennifer Sano-Franchini, and Angela M. Haas. 2018. "Interfacing Cultural Rhetorics: A History and a Call." *Rhetoric Review* 37 (2): 139–154.

Cox, Matthew B., Elise Dixon, Katie Manthey, Maria Novotny, Rachel Robinson, and Trixie G. Smith. 2021. "Embodiment, Relationality, and Constellation: A Cultural Rhetorics Story of Doctoral Writing." *Re-Imagining Doctoral Writing*, edited by Cecile Badenhorst, Brittany Amell, and James Burford, 145–166. Fort Collins, CO: The WAC Clearinghouse/ University Press of Colorado.

De La Trinidad, Maritza, Stephanie Alvarez, Joy Esquierdo, and Francisco Guajardo. 2021. "Historias Americanas: Implementing Mexican American Studies in K–12 Social Studies Curriculum in the Rio Grande Valley." *Association of Mexican American Educators Journal* 15 (2): 36–63.

Edwards, Dustin W. 2020. "Digital Rhetoric on a Damaged Planet: Storying Digital Damage as Inventive Response to the Anthropocene." *Rhetoric Review* 39 (1): 59–72.

Guajardo, Francisco, and Miguel Guajardo. 2013. "The Power of Plática." *Reflections* 13 (1): 159–164.

Kerschbaum, Stephanie L. 2015. "Anecdotal Relations: On Orienting to Disability in the Composition Classroom." *Composition Forum*, no. 32. https://compositionforum.com/issue/32/anecdotal-relations.php.

Medina-López, Kelly. 2018. "Rasquache Rhetorics: A Cultural Rhetorics Sensibility." *Constellations: A Cultural Rhetorics Publishing Space* 1 (1): 1–20. http://constell8cr.com/issue-1/rasquache-rhetorics-a-cultural-rhetorics-sensibility/.

Pough, Gwendolyn D. 2011. "2011 CCCC Chair's Address: It's Bigger than Comp/Rhet: Contested and Undisciplined." *College Composition and Communication* 63 (2): 301–313.

Powell, Malea. 2012. "2012 CCCC Chair's Address: Stories Take Place: A Performance in One Act." *College Composition and Communication* 64 (2): 383–406.

Powell, Malea, Daisy Levy, Andrea Riley-Mukavetz, Marilee Brooks-Gillies, Maria Novotny, and Jennifer Fisch-Ferguson. 2014. "Our Story Begins Here: Constellating Cultural Rhetorics." *Enculturation: A Journal of Rhetoric, Writing, and Culture* 25: 1–28.

Riley-Mukavetz, Andrea M. 2014. "Towards a Cultural Rhetorics Methodology: Making Research Matter with Multi-Generational Women from the Little Traverse Bay Band." *Journal of Rhetoric, Professional Communication, and Globalization* 5 (1): 109–125.

Index

Page numbers followed by *n* indicate an endnote.

absence of interdisciplinarity, 64
academia: belonging, sense of, 183; classism, 132, 135, 141, 145; discriminatory practices, 143–44; elitism, 53; gatekeeping, 184–85; microaggression, 148–49; navigation, 138; politics of, 107; tokenism, 179–80
academic disciplines, 4, 5, 16
academic guilt, 134, 140
academic identity. *See* professional identity
academic lineage, graduate students of color, 223
academic space: diversity, 145, 168, 208; literacy, 186–87; navigating, 179; social uplift, 191–93; support systems, 142–44, 200
accountability, predominantly white institutions, 208
accumulative conceptual frameworks, 55
Ackerman, John M., 4, 9, 23, 47, 62
acknowledgement of power dynamics in writing studies, 235
acquiescence, claiming disciplinarity, 136, 138, 139
activism, writing studies, 142
adaptation, writing studies programs, 79
addressing privilege, 235
adjunct faculty: administrative positions, 88; claiming disciplinarity, 107–8; composition and rhetoric studies, 109; first-year writing courses, 108; job security, 10–11, 109–12, 119–21; professional development, 127–29; professional identity, 121
Adler-Kassner, Linda, 63, 64
administrative positions: adjunct faculty, 88; department formation, 45; disciplinary status, 106; first-year writing studies, 91; higher education, 13; structures, 77; tenured faculty, 91; writing studies, 86
advising, independent writing groups, 113
affect theory, 16, 196, 200–2, 205
affective labor, 132–34; assimilation, 136; job market, 141–42; marginalized students, 143–44; pathways into the discipline, 142; queer scholars, 135; scholars of color, 170
affirmative action, 28
African American narrative, 212–13
African American students, 59*n*14
African American Vernacular English (AAVE), 183, 217, 223*n*1
agency: critical literacy, 192; first-year composition, 13; material feminism, 198; writing program administrators (WPAs), 72–73
Agnew, Eleanor, 53, 59*n*9
Ahmed, Sara, 78, 152, 153
Alaimo, Stacy, 198

240 : INDEX

alienation, scholars of color, 159–60n1
allegiances, 15, 30–31, 165
allyship, 221
alphabetic literacy, 211, 219
alternatives, hegemonic systems, 79
Ann Arbor public school systems, 216, 217, 218
Anson, Chris, 55–56
anti-Black linguistic racism, 223n1
antiracism practices: coding literacy, 222; first-year writing composition, 13; graduate programs, 145; open admissions policies, 54; public school systems, 217–18; role of literacy, 208–9; scholarship, 142–43; training writing instructors, 223; writing studies, 10, 221
Anzaldúa, Gloria, 149, 156, 160n6, 160n7
Article 11G, 110–14, 115–16, 119–20, 122, 123–25
assimilation: Black students in the classroom, 217–18; claiming disciplinarity, 136; graduate students of color, 149; job security, 234; professional identity, 234; professionalism, 136; scholars of color, 160n7; tokenism, 167; whiteness, 133, 134; writing studies, 145
assistant professorship appointments, 129n2
Association of Pennsylvania State College and University Faculties (APSCUF), 109
authoritative discourse, 217
autobiography, 16
autoethnography, 16, 77, 79–80, 105
autonomy, 126

Baker, Edith, 30, 31, 32, 36
Bakhtin, Mikhail M., 217
Ball, Cheryl, 210
Banks, Adam J., 4, 63, 219
Banks, William P., 144
Barad, Karan, 202
Bartholomae, David, 55
basic writing courses, 52, 53
Bazerman, Charles, 5–6
becoming, process of, 6–7, 104, 222
being claimed, 114; writing studies, 105
Bell, Derrick, 214
belonging, sense of, 5, 8, 10, 36, 138, 145
Bernhardt, Stephen, 29, 37, 40
BFGS. See Black female graduate students
bias, coding literacy, 220
Bibb, Maria, 220
Billingsley, Khadeidra, 10, 15, 16, 134, 165, 192, 229, 231
binaristic thinking, 79
Bizzaro, Resa Crane, 133, 134, 136, 139, 140
Bizzell, Patricia, 149

Black Arts Movement, 223n2
Black female graduate students (BFGSs), 168; emotional labor, 169, 174–77; intellectual labor, 176–77, 179; research participants, 177; token labor, 178, 180. See also scholars of color
Black, Indigenous, and Other People of Color (BIPOC): academic identities, 208; deficit thinking, 221; pathways into writing, 221; professional access to intellectual space, 208; research agendas, 222–23; student literacy, 29; teaching writing, 209. See also scholars of color
Black intellectuals: assimilation, 217–18; community responsibility, 212; literacy theories, 218–21; philosophy of education, 212–13, 214, 216, 219–20
Black linguistic practice: politicization, 217; public school systems, 216, 217; rhetoric and composition studies, 218; writing studies, 209. See also scholars of color
Black representation: digital literacy, 221–22; doctoral programs, 168, 170–71, 212–15; rhetoric and composition studies, 215, 216–17
body, marginalized, 171–72
Bollig, Chase, 137
Bootstraps: From an American Academic of Color, 184
border tracing, 5–6
borderlands, 73
boundaries between disciplines, 64
Bowen, Lauren Marshall, 12, 16, 21, 44, 45, 58n1, 64, 98, 109
Brady, Ann, 153
Brannon, Lil, 36, 37, 39
Brereton, John, 41
Bridges, D'Angelo, 191
Bridgewater, Matthew, 71
Brooks, Ron, 11, 13, 16, 21, 45, 72, 77, 79, 105, 111, 233
Brown, Haillie Quinn, 4
Brown, Mike, 209, 212, 213, 214
budgets, 62
burnout, 134
Burton, Larry, 87
business writing courses, 70, 96
bylaws, 88
Byrd, Antonio, 16, 24, 165

calcium carbonate, 160n5
Cano, José, 155
cap on non-tenure-track faculty, 110–11

career choices, 137–38, 208
CCCC (Conference on College Composition and Communication), 25, 36, 54, 79
Cedillo, Christina V., 16
Ceraso, Steph, 144
certificate programs for professional development, 128
challenges, first-year composition, 41n1, 62
changing disciplines, 3–4, 38
Chaput, Catherine, 204
Chávez, Karma, 171
Chicanx scholars: lived experiences, 156, 185–86; otherness, 155; professional identity, 155–56; writing studies, 157
CIP (Classification of Instructional Programs) codes, 23, 63
circulating norms, 78
citation practices, 145, 223
City University of New York (CUNY), 12, 28, 29, 52–54
Civil Rights Movement, 211
claiming disciplinarity, 73, 114; acquiescence, 136, 138, 139; administration positions, 67–68; assimilation, 136; class-based, 185; creative writing, 106; cultural and historical context, 24; doctoral programs, 48–50; economic circumstances, 45, 98; historical observation, 57; identity, 30; institutional circumstances, 45; literacy, 192–93; literature/composition divide, 38, 88–90; lived experiences, 153–54, 185, 187–90; material labor, 129; navigation, 136, 137, 138; negotiation, 103, 155; origin stories, 44–45; personal circumstances, 31–32; political circumstances, 45, 98; race-based, 185; resilience, 13–14, 103, 152–54; self-reflection, 176; slow agency, 68; specializations, 110; strategy, 26; tactical, 26; teaching of writing, 32–33; writing studies, 5, 8, 24, 38, 62, 77, 105, 108, 126, 129, 145, 201
Clary-Lemon, Jennifer, 200–201
Classical Rhetoric for the Modern Scholar, 48
Classification of Instructional Programs (CIP) codes, 23, 63
classism: academia, 29, 132, 135, 141, 145; class dissonance, 138; disciplinarity, 185; identity, 184; professionalism, 133
classroom management, 219
Claycomb, Ryan M., 70
climate change, 198, 235
close readings, 55
co-optation, positionality, 191

coalition-building, 36
code-meshing, 156, 182, 193, 216, 218
code-switching, 155, 187, 187, 218
coding literacy, 215, 219–22
collaboration initiatives: autoethnography, 77, 79–80, 105; cross-disciplinary studies, 36–37, 93; curriculum development, 93; department formation, 88, 98; peer networks, 142; writing studies programs, 35, 77; collective bargaining agreements, 109
collective reality, 203, 204
College Composition and Communication, 67
College Research Paper, 64
colleges and universities: effecting change, 21–22; financial challenges, 28; labor models, 78; open admissions policies, 12, 54–55
Collins, Patricia Hill, 214
colonization of intellectual space, 208
committee memberships: department formation, 84–85, 92; instructional specialist faculty, 85, 88, 91–92; tenure-track faculty, 84–85, 116
community colleges, 108–9
community literacy, 58n2, 184
community responsibility, 212
composition, rhetoric and writing studies (CRW), 86; adjunct instructors, 105, 108–9; claiming the discipline, 5; creative writing, 120, 121; integration with literature studies, 58n3; peer workshops, 113; requirements, 30; writing pedagogy, 126–27. *See also* rhetoric and writing studies
Composition, Rhetoric, and Disciplinarity, 45, 51, 58n7, 63
composition/literature divide, 36–38, 88–90
computer programming. *See* coding literacy
Conference on College Composition and Communication (CCCC), 25, 36, 54, 79
connection through language, 36, 56, 202, 230, 235, 237n2
consensus leadership in writing studies, 9, 86
contemporary disciplinarity, 45
content/outcomes articulations, 73n1
Contingency, Exploitation, and Solidarity: Labor and Action in English Composition, 11
Contingency, Solidarity, and Community Building: Principles for Converting Contingent to Tenure Track, 129n1
contingent faculty, 97, 98
conversion of long-time colleagues, 112
conversion process, 112, 115, 116, 117–22, 123, 125–26

Cooper, Anna Julia, 219
Corbett, Edward P.J., 48
core curriculum committees, 69
corporeal analysis of course material, 235
correctness, writing pedagogy, 197
counterstory, 52
course offerings, 49–50, 93
COVID-19 pandemic, 11
Cox, Matthew B., 144
creating space, 230–31
creation of knowledge, 195
creative nonfiction writing, 108, 210
creative writers: claiming the writing studies discipline, 13–14, 106; professional identity, 114–15, 123–24; teaching of writing, 121
creative writing studies: courses, 111, 112; degrees, 136; disciplinary identity, 3–4, 6–7, 126; faculty, 105, 127; job security, 107–8, 126; multidisciplinary, 7; PhDs, 128–29; students of color, 151; terminal degrees, 128–29; undergraduate major, 151; whiteness, 151; workshops, 115
critical code studies, 220–21
critical language awareness, 215, 218
critical literacy, 192–93
critical race theory, 218
cross-disciplinarity: claiming, 62; collaboration with other departments, 93; doctoral programs, 49–50; learning, 234; scholarship, 38, 63–64, 195; teaching of writing, 113; writing studies programs, 64
cross-generational engagement, 24
Crow, Angela, 87
Crowley, Sharon, 4
Cruising Utopia, 85
cultural analysis of course material, 41, 211, 213, 219, 220, 230, 235
cultural diversity, 24, 191, 222–23
cultural studies/empiricism divide, 38
CUNY (City University of New York), 12, 28, 29, 52–54
current-traditional pedagogy, 28, 30, 44–45
current trends in writing studies, 97
curricular revisions, 69; first-year composition, 65–66, 96–98; rhetoric and composition studies, 29–30; shared decision-making, 97–98
curriculum development, 13; coding literacy, 219–20; collaboration with other departments, 93; core committees, 69, 84–85; creative writing instructors, 127; deficit thinking, 92; digital literacy, 210, 219;

English departments, 80; first-year writing courses, 30, 62; graduate programs, 23; inclusion, 98–99; norms, 90–91; plurilingualism, 183; standardization, 66–68; writing across the curriculum (WAC)/writing in the disciplines (WID), 62, 69–70; writing studies, 23, 73n1, 77, 81, 94, 95
CWPA (Writing Program Administrators, Council of), 35

Dadas, Caroline, 11, 13, 16, 21, 45, 72, 76, 77, 79, 105, 140, 143, 144, 233
Dallas, Phyllis Surrency, 59n9
Daniel, James Rushing, 144–45
Dartmouth Seminar (1966), 216
de Certeau, Michel, 26
decolonization of writing studies programs, 159
Decolonizing Rhetoric and Composition: New Latinx Keywords for Theory and Pedagogy, 159
deconstructed myths, 58n8
deficit thinking, 53, 92, 216, 217, 221
democratization, coding literacy, 221–22
demographics, graduate students, 59n15, 59n16, 140
Denny, Harry, 139
department formation: administrative priorities, 45; bylaws, 88; challenges, 93; collaboration, 88, 98; committees, 85, 92; contingent faculty, 97; disciplinary identity, 46–47; enrollments, 81; equity, 98; funding, 45, 47; inclusion, 98; interdepartmental relations, 90, 93; multidisciplinary structure, 7; negotiation, 45–46; self-definition, 87; shared decision-making, 97–98; state requirements, 45; union contracts, 47, 106; visibility, 46; writing studies programs, 58n5, 58n6, 58n7, 80–81, 86–87, 106
descriptive writing, 208
Dew, Debra Frank, 58n3
digital composing practices, 210
digital literacy, 209–11, 218–21
disability rhetorics, 140, 235
disciplinarity, claiming: acquiescence, 136, 138, 139; administration positions, 64, 67–68, 106; allegiances, 30–31; arriving to, 11; assimilation, 136; belonging, 36, 145; changing attitudes, 3–4, 38; class-based claiming, 85; coalition-building, 36; consensus, 9; creative writing, 126; cross-disciplinary learning, 234; department

formation, 46–47; disciplinary identification, 187; doctoral study, 48–50; economic circumstances, 45, 98; epiphany narratives, 103; equity, 15, 144; frameworks, 55; institutional circumstances, 27–28, 45, 47, 95, 144–45; interview participants, 25, 26; literature/composition divide, 38, 88–90; lived experiences, 153–54, 185, 187–90; navigation, 132, 136, 137, 138; negotiation, 103, 155; new disciplinarity, 73; origin stories, 26, 29, 44–45; political circumstances, 45, 98; race-based claiming, 185; research participants, 27–28, 38–39; resilience, 103, 152–53; rhetoric and composition studies, 62, 201; self-reflection, 176; slow agency, 68; specializations, 39, 40, 110; static disciplinarity, 73; storying, 12; strategic claims, 26; tactical claims, 26, 29; taxonomy, 9; value, 47; whiteness, 4, 16; writing studies instructors, 8, 14, 23–26, 37, 126, 145

disciplinary connections, first-year composition, 41n1

disciplinary knowledge, 38, 56–57, 184

disciplinary legitimacy, 35, 45, 208

disciplinary purity, 98

discrimination, scholars of color, 143–45, 149–52, 169, 190–91, 229, 235

discursive space, 221, 237n2

disrupting norms, 78

dissertations, 50

diversity, equity, inclusion (DEI): citation practices, 223; classrooms, 143; college student populations, 31; community colleges, 108; curriculum development, 98–99; disciplinarity, 40–41, 144; English departments, 110; fellowships, 232; frameworks, 223; linguistic practices, 15; literacy practices, 15, 193; multicultural narratives, 230, 233; open access admissions, 234; pathways into the discipline, 142; predominantly white institutions (PWIs), 208; research participants, 25, 59n15, 59n16, 209; supportive spaces, 143–45; teaching practices, 209; textual evaluation, 184; writing studies, 38–39, 98, 109, 230

Doctoral Consortium, 9

doctoral study/programs, 58n9, 128–29; adjunct faculty, 128; Black representation, 170–71, 212–15; claiming disciplinarity, 48–50; financial labor, 49, 140–41; intellectual labor, 176–77; job market, 31, 140–41;

personal circumstance, 49; rhetoric and composition studies, 27; specializations, 49; students of color, 169–70; writing studies, 9, 59n10

dominant narrative, 133, 167–68, 217, 220, 221, 232

Donawerth, Jane, 34, 35

Downs, Doug, 5

The Dream of a Common Language, 125

dress codes, 143

Driscoll, Dana Lynn, 134, 140

dropout. *See* student retention rates

Du Bois, W.E.B., 4, 219, 220

Duffey, Suellynn, 16, 12, 21, 24, 95, 234

Dunbar, Paul Laurence, 217

dynamis (potentiality), 203

Dyson, Michael Eric, 212

Early Performance Feedback, 71

Eaton, Adrienne E., 173

Ebonics. *See* African American Vernacular English (AAVE)

economic circumstances for claiming, 45, 98

educational theory, 212–13, 219–20

effecting change at the university level, 21–22

elasticity, 73

elitism in academia, 53

embodied knowledge. *See* lived experiences

emerging field of writing studies, 29, 63

Emig, Janet, 59n15

emotional labor: academic guilt, 140; burnout, 140; career choices, 137–38; imposter syndrome, 140, 191; professionalism, 134–35; students of color, 158, 232; teacher-scholars, 103–4, 141–43, 168, 175–77, 186; tokenism, 134, 169, 192; emotional trauma, 210–11

empiricism/cultural studies divide, 38

emplacement, 230, 235

employment. *See* job security

enactment, 230, 235

endurance strategy, 179–80

energeia (actuality), 203–4

English Composition Board (University of Michigan), 29

English degrees, 136

English departments: curriculum development, 80; disciplinary purity, 98; diversity, lack of, 110; embedded writing studies, 80; hierarchy, 99; instructional specialist positions, 82–83; literature/writing divide, 37, 81, 88–90; otherness, 37; separation from writing studies programs, 76, 79–80,

94–95, 99n1; social justice initiatives, 113–14; writing studies, 207–8
English for the Children initiative, 183
English language learners (ELLs), 183
enrollment in higher education, 28, 52–54, 80, 81, 140
environment, affect, 16
epiphany claiming narratives, 103
episodic disciplinarity, 55
epistemological theorizing, 51, 52
equity. *See* diversity, equity, inclusion (DEI)
Errors and Expectations, 53
Ersheid, Alison, 11, 13, 103, 234
established norms, 78
ethnicity as barrier in writing studies, 159
ethnography, 11, 105
Everett, Justin, 77, 86
evolution of writing studies programs, 27, 38–39
exclusion, professionalism, 139, 158–59
expectations, scholars of color, 138, 168, 186
exploitation, influence on writing studies, 235

faculty: antiracist practices, 151, 209, 223; collective bargaining agreements, 109; contingent, 97; financial precarity, 141–42; health issues, 158; job security, 31–32, 207; lived experiences, 51–52, 57; non-tenure-track, 105; personal identity, 37; professional development, 68, 69, 97, 219; professional identity, 114–15, 123–24; retirement, 24–26, 44; self-identity, 25; supportive spaces, 143–44; union contracts, 77, 106. *See also* instructional specialist faculty; non-tenure-track faculty; tenure-track faculty; teaching of writing
failure, concept of, 78, 99
fair labor practices, 77
Feinstein, Bruce, 71
fellowships, 232
female intersubjectivity, 195
female scholars of color: emotional labor, 134; endurance strategy, 179–80; horizontal mentoring, 144; imposter syndrome, 160n2; identity, 14; intersectionality, 152; research participants, 25; resilience, 149, 153–54, 159; rhetoric and composition studies, 198–99; tokenism, 179–80
feminism, 157
feminized disciplinarity, 36
Ferguson, Missouri, 16, 212, 214
fiction writing, 207
Field, Laura, 11, 13, 21, 45, 72, 77, 79, 233

Field of Dreams: Independent Writing Programs and the Future of Writing Studies, 58n6, 58–59n9, 87
financial labor: challenges, 28; doctoral candidates, 49, 141; graduate students, 136, 138, 140; influence on career choices, 132, 137–38; non-tenure-track faculty, 135–36; part-time faculty, 141–42; pathways into the discipline, 132; scholars of color, 172, 173, 174, 175–76; student debt, 144–45
Fine, Michelle, 31
first-generation students, 49, 53, 70–72, 73n1, 231
first-year writing (FYW) programs: adjunct instructors, 107–108; administrative positions, 91; antiracist practices, 13; assessments, 71, 72; challenges of, 62; community colleges, 108–9; current-traditionalist, 30; curriculum development, 62, 65–66, 96–98; digital composing, 210; disciplinarity, 41n1, 65–66; expectations, 64; inclusion, 98; mentoring, 68; pedagogy, 108; peer workshops, 113; politics, 39; relevance to students, 30; research papers, 61; slow agency, 13; state requirements, 72; student retention rates, 71; teaching of writing, 121; text remediation, 210–11; writing studies, 70
5/5 workloads, 110
Fleming, David, 30, 55–56
flexibility in writing studies programs, 59n11, 79, 96
Flynn, Elizabeth, 153
foundation grants, 29
4/4 teaching workload, 81, 110
frameworks for cultural diversity, 222–23
full-time faculty, relationship with tenure-track faculty, 78
funding for professional development, 34, 45, 47, 127
future of writing studies, 233–34

Gannett, Cinthia, 41
García de Müeller, Genevieve, 134
Garrett, Nathan, 71
Garvey, Marcus, 219
gatekeeping in academia, 133, 184–85
Gates, Jr., Henry Louis, 213
gender identity, 58n4, 184
gender studies, 77
general education curriculum, 69, 73n1, 96, 110
generational subsets, 58n1
geographic location, claiming a discipline, 159

George Mason University, 107
George, Diana, 69
Georgia Southern, 47
Gere, Anne Ruggles, 36, 73
Gilyard, Keith, 4, 184, 216
Gladwell, Malcolm, 216
Glenn, Cheryl, 160n7, 184, 190, 192
Glover, S. Tay, 168, 175, 177, 178
Google Docs comment feature, 79
graduate programs: antiracist practices, 145; development, 23; diversification, 232; rhetoric and composition studies, 68; teaching assistants, 107; writing studies, 23, 99n1
graduate students of color: assimilation, 149; claiming space, 168; predominantly white universities, 15; research agendas, 222–23; financial labor, 135, 138, 140; first-generation, 231; resilience, 149, 159; supportive spaces, 143–44; tokenism, 232–33; workload, 139–40
Graff, Harvey J., 191, 220
grants, 29
Green, Dari, 173
Grollman, Eric Anthony, 176
Gullah dialect, 217

hacker culture, 221
Hairston, Maxine, 36–37
Halberstam, Jack, 78, 85
Hall, Stuart, 189
Hanganu-Bresch, Christina, 77, 86
happy accident narrative, 26
Hariman, Robert, 133
Harlem Renaissance, 220, 223n2
Harris, Muriel, 56
Haswell, Rich, 38
Haviland, Carol, 32, 33
Hawhee, Debra, 160n8
Hawk, Byron, 56
health issues, faculty, 158, 171–72, 176
Heath, Shirley, 187
hegemonic systems, 78, 79
Heilker, Paul, 5
Hekman, Susan J., 198
Herzberg, Bruce, 149
hidden curriculum, 133, 143–44
hierarchy in academia, 99, 187, 235
higher education: administration, 13; antiracist practices, 223; Black representation, 213; classism, 29, 145; cultural climate, 41; diversity, equity, inclusion (DEI), 232–33; employment, 11; enrollment, 28; first-generation students, 73n1; institutional climate, 41; linguistic racism, 29; neoliberal policies, 11; open admissions policies, 234; political climate, 41; tokenism, 167
Hillard III, A. G., 212
hiring ethics, 113, 145
Hispanic teacher-scholars, 148
Hispanic-Serving Institution (HSI) status, 70, 98–99, 73n1
historical context of writing studies, 12, 23, 24, 27
historical narratives, 55, 57
historiographic studies, 11
"The History of Open Admissions and Remedial Education at the City University of New York," 59n13
Hochschild, Arlie, 134
hooks, bell, 179
horizontal mentoring, 144
Horner, Bruce, 55
Horning, Alice, 58n3
Howells, William Dean, 217
Hull, Brittany, 143
human experience, 198–99
hybridity, scholarship, 195
hypervisibility, tokenism, 167, 170–71

idea-building, 142
identity. See personal identity; professional identity
idiolects, 182
illiteracy, 220
imposter syndrome, 134, 140, 160n2, 165, 175, 178, 191, 231, 232
inclusion. See diversity, equity, inclusion (DEI)
independent writing studies programs, 58n5, 58n6, 58n7, 113, 209
Indigenous history, 58n8, 201
individual identity. See personal identity
inequity, computer programming in public education, 219
Ingraham, Chris, 204
Inoue, Asao B., 143
institutional disciplinarity, 41, 45, 47, 95, 144–45
institutional ethnography, 16, 105
institutional racism, 183–84, 214
institutional recognition of writing studies programs, 35, 45, 93
instructional specialist faculty: career trajectories, 208; committee membership, 84–85, 88, 91–92; creative writing degrees, 58n9; labor conditions, 105; literature

degrees, 58n9; marginalization, 83–84; rhetoric and composition degrees, 59n9, 105; service positions, 39, 81–83, 89; workload, 25, 76, 78, 81, 89, 91–92, 107; writing studies programs, 99n2. *See also* faculty; teaching of writing
instructional strategies, 219
instructors. *See* faculty; instructional specialist faculty; teaching of writing
intellectual labor, 34–35, 176–77, 208
interconnectedness, cross-disciplinary studies, 33, 36–37, 204
interdepartmental relations, 90, 93
interdisciplinary connectedness, 5–6, 36, 41n1, 63–64, 77, 218
Internet, 209
intersectionality, professional identity, 24, 152
interview participants. *See* research participants
intuition, effect on writing, 196
invisibility/visibility, 9, 13

job market, 140–42
job security: adjunct faculty, 10–11, 111–12, 119–21; assimilation, 234; creative writing instructors, 107–8, 126; doctorate-holding job seekers, 31; instructional specialist positions, 82–83; non-tenure-track faculty, 10–11, 114, 123, 135–36, 139–40; teachers of color, 59n14; teaching of writing, 31–32, 207; technical writing degrees, 32, 136–37; tenure-track faculty, 114, 116, 122, 124–26; terminal-degree-holding faculty, 119; writing studies programs, 106, 107
Johnson, Cynthia, 14, 32, 103–4, 149, 168, 208
Johnson, Jessica Marie, 214
Journet, Debra, 31–32, 34

Kahn, Seth, 11, 14, 16, 103, 234
kairos, 12, 13, 47, 77
Kanter, Rosabeth Moss, 170, 171
Kates, Susan, 4
Kennedy, George A., 203
Kerschbaum, Stephanie, 235
key words for writing studies, 5, 63
Kimball, Elizabeth, 54–55, 57, 59n13
Kinneavy, James, 31
Kirsch, Gesa, 153
Kitchen Cooks, Plate Twirlers, and Troubadours: Writing Program Administrators Tell Their Stories, 69
knowledge-making methodologies, 41, 57, 63, 195, 200, 201, 202

Koenigsburg, Lisa, 11, 13, 103, 234
Kristensen, Randi Gray, 70
Kynard, Carmen, 174

labor models, 77, 78, 99, 105, 179
Labov, William, 217
LaFrance, Michelle, 105
Lalicker, William B., 11, 129n1
language: barriers in writing studies, 159, 183–84; connection, 202; constructs, 204; human experience, 198–99; knowledge-making, 200, 201, 202; limitations of, 205; lived experiences, 165; love of, 196–97; power of, 15–16, 205; preservation, 221; white racial habitus, 208, 209, 217
Lau v. Nichols (1974), 183
Lauer, Claire, 8
Lauer, Janice, 27, 36, 37, 38–39
Lazerson, Marvin, 31
Leigh, S. Rebecca, 134, 140
lesson plans, 209–11
LGBTQ scholars, 140, 141
Lindemann, Erika, 26, 35, 39
linear frameworks, 55
linguistic pluralism, 183
linguistic racism, 15, 16, 29, 217, 218, 223n1
listservs, 25
Literacy and Racial Justice: The Politics of Learning after Brown v. Board of Education, 59n14
literacy: Black intellectuals, 219–21; claiming, 192–93; code-meshing, 182; coding literacy, 221–22; community, 184; crisis, 27–28; cultural hegemony, 191; equity, 15, 193; hierarchy, 187; idiolects, 182; institutional, 184; literacy-focused programs, 29; myths, 158; potentialities, 193; role in antiracism, 208–9; social uplift, 191–92; spaces of, 186–87; standard English, 223n1; stratification, 187; subjectivity, 211; white racial habitus, 208, 209
literary studies: changing disciplines, 3–4, 46; English majors, 80; integration with composition and rhetoric studies, 58n3; literature degrees, 136; literature/composition divide, 36–38, 88–90, 113; multidisciplinary, 7; writing studies faculty, 6–7, 28, 32–33, 35, 40, 44–46, 112
lived experiences, 57, 157, 234; claiming disciplinarity, 151–54, 185, 187–90; coding literacy, 211, 219; effecting change at the university level, 21–22; epistemology, 46, 51–52; female, 195; language, power of, 165; marginalized rhetorical tactics, 209;

meaning-making, 229; otherness, 235; professional identity, 30, 184–85; scholars of color, 15, 156, 185–86, 219, 222, 229; storying, 229; teaching of writing, 11, 51–52, 57; tokenism in academia, 168, 179–80; writing studies, 15, 16
Lives on the Boundary: The Struggles and Achievements of America's Underprepared, 184
Living a Feminist Life, 153
Lu, Min-Zhan, 55
Lynch-Biniek, Amy, 11, 129n1

Mack, Nancy, 33, 37, 38–39
Maid, Barry, 46
mainstream writing studies, 221
Majors and Minors, 217
Malencyzk, Rita, 71–72
Malinche, 155
The Managed Heart: Commercialization of Human Feeling, 134
management resistance, 117–22
Manthey, Katie, 143
"Mapping the Terrain of Tracks and Stream," 59n12
marginalized students, 29, 37, 83–84, 143–44, 208, 209, 223, 229, 230
Marino, Mark, 220
market-driven writing studies programs, 86, 99, 114
marketing writing, 107
Marshall Bowen, Lauren, 233–34
Martin, Joe, 32–33
master's degrees, 58n9, 59n10, 99n1, 107–9, 128–29, 129n2
Mateas, Michael, 220
material analysis of course content, 235
material interaction, 16, 195, 197, 198–200, 202, 204, 205
material labor terms, 14, 129
Matsuda, Paul, 216
Mays, John, 184
McGuire Memorandum, 110–12
Mckoy, Temptaous, 143
McLaughlin, Margaret, 53
McSpadden, Lezley, 212, 213
McVeigh, Maureen, 11, 13, 24, 103, 234
meaning-making, 208, 219, 229, 230, 234–35
Medina, Cruz, 155
mental health, 152, 172, 175–76, 232
mentoring, 28–29, 34, 68, 144
merit fellowships, 232
messiness in queer methodologies, 99
mestizaje, trope of, 155

methodologies, 16, 24–25, 41, 78
Micciche, Laura, 13, 68, 72–73
Michigan, University of, 29
microaggressions, 148–49, 151–52, 160, 190–91
Miller, Carolyn R., 137, 143–44
Miller, Liz, 139
Miller, Susan J., 48
A Mindfield of Dreams, 77
minoritized students, 52, 53, 54, 191
Mlynarczyk, Rebecca, 28–29, 32, 33, 37, 39
Montclair State University, 76, 79, 98–99
Montfort, Nick, 220
Moore III, James L., 177
Moss Kanter, Rosabeth, 167
multicultural narratives, 10, 230, 233
multidisciplinary writing departments, 7
multimodal composition. *See* digital literacy
Muñoz, José Esteban, 85, 160n7
Murphy, Michael, 128

narrative diversity, 52, 221, 230, 233
national searches, 112
navigation of academic spaces, 136, 137, 138, 179
Neff, Joyce, 31, 35
negotiation, claiming disciplinarity, 45–46, 49, 54, 103, 106, 155
neoliberal policies, 11
networks of care, 143–44
New Criticism, 55
new disciplinarity, 73
new materialism, 200–201
Niemann, Yolanda Flores, 160n2, 167, 169
non-tenure-track faculty: cap, 110–11; collective bargaining agreements, 109; conversion process, 115, 116, 117, 122, 123; financial labor, 135–36; instructor workload, 110; job security, 10–11, 105, 114, 123, 139–40; material labor as pathway to tenure, 14; McGuire Memorandum, 110–11; professional development, 127–29; teaching positions, 111–12; union contracts, 106; workload, 110
nonfiction essays, 210
nonlinear writing style, 149
norms in academia, 78, 90–91, 139

O'Neil, Peggy, 87
Oakland Resolution against Ebonics, 223n1
object-oriented ontology, 198–99, 201
objectivity in writing frameworks, 214
Olson, Gary, 5
ontological approach, 197–98

open admissions policies, 12, 28, 52–55, 59*n*13, 234
opioid epidemic, rhetoric of race, 189
opportunities in first-year composition, 41*n*1
oral narratives, 212–13
Ore, Ersula, 191, 213
origin stories, 7, 8, 23, 26, 34, 44–45
otherness, 37, 46, 58*n*4, 151–52, 155, 159–60*n*1, 160*n*2, 231, 235

parent scholars, 31
part-time faculty, 91–92, 118–19
pathways into the discipline, 48–49, 142, 144–45, 211, 218, 221
Pearson, Nancy, 11, 13, 103, 234
pedagogical antiracist initiatives, 218
pedagogical autonomy, 126
pedagogical racism, 29
pedagogical shift, 44–45
pedagogy versus theory practice, 5, 40–41
peer networks, 35, 113, 142
Pennsylvania, 129*n*2
performance expectations, 71, 168
Perl, Sondra, 56, 59*n*15
permeability of writing studies, 10
Perry, Theresa, 212, 216, 220
Perryman-Clark, Staci, 179
persistence rates of students, 70–72
personal circumstances, claiming the discipline, 15–16, 30–32, 41, 49
personal identity, 5, 15–16, 31, 157, 198–99
personal reflection, 16
personal statement in research, 214–15
PHD to Ph.D.: How Education Saved My Life, 184
Phelps, Louise, 4, 9, 47, 62, 69
philosophy of education, 212–13
physical labor, 168
Pinkert, Laurie A., 12, 16, 21, 44, 45, 58*n*1, 64, 98, 109, 233–34
pivotal strategy, 7–8, 95, 143–44
placement systems, 53
plática, 230, 237*n*2
plurilingualism, 183, 184
poetry, 108–9
points of entry, 33–34
polarization, 167, 175–76
policy changes for writing studies programs, 105–6
political analysis of course material, 235
political circumstances for claiming disciplinarity, 39, 45, 98
political writing courses, 96
politics in writing studies programs, 105–7

positionality: co-optation, 191; knowledge, 195; language, 205; personal circumstances, 31; research methods, 233–34; scholars of color, 191; scholarship, 195; teacher-scholars, 103; white privilege, 132; writing studies programs, 14, 158, 235
posthumanism, 197–200, 204
power dynamics, 235
power of language, 205, 230
practicality in writing courses, 96
pre-disciplinary writing studies programs, 26, 27
predominantly white institutions (PWIs), 15, 167, 168, 191, 208
Prendergast, Catherine, 59*n*14
print-only, 210–11
privilege, 14, 141–42, 235
process research, 55–56
professional development: adjunct faculty, 127–29; Conference on College Composition and Communication (CCCC), 25, 36, 54, 79; certificate programs, 128; creative writing studies, 115; CWPA conferences, 35; faculty, 97, 98, 127–29; funding, 34, 127; teaching of writing, 68; WAC/WID, 69, 70
professional identity: adjunct faculty, 121; antiracist practices, 223; assimilation, 234; claiming disciplinarity, 30, 49; creative writers, 114–15, 123–24; cultural conditions, 24; curricular revisions, 96; disciplinarity, 23, 24, 30, 208; feminist scholars, 14; gender, 58*n*4; influences, on, 235; language, importance of, 15–16; lived experiences, 30, 184–85; norms, 139; resilience, 104; scholars of color, 58*n*4, 155–56, 170–71, 176–77, 179, 187, 208, 222, 232; shaped by language, 15–16; social conditions, 24; standard conventions, 208; teacher-scholars, 165; tokenism, 168; training, 109; writing studies, 7–8, 24, 80
professional labor, 168
professional organizations, 143–44
professionalism: assimilation, 136; challenges, 142; classism, 133; deconstruction of, 14; diversity, 145; emotional labor, 134; exclusion, 139; expectations, 138; graduate students, 223; hidden curriculum, 133; job market, 141–42; peer networks, 35; systematization of knowledge, 133; tenure-track faculty, 141–42; upward mobility, 133, 139; white privilege, 136; writing studies, 132
professorships, 129*n*2
program-building, 11, 21–22

Progressive Era, 220
project management, 107
public and professional writing (PPW), 79, 94
public education systems, 216–19

qualitative methodologies, 16, 71, 79–80
queer affective labor, 135
queer collaborative autoethnography, 105
Queer Phenomenology, 78
queer theory, 13, 78, 79, 92, 96, 97, 99, 143–45, 233
quiet labor, 145

race and technology, 215, 219–22
race identity, 58n4, 184, 208, 209
race-based initiatives, 135, 185
racial diversity, lack of, 25
racial integration, 53
racial politics, 5
racially-marginalized literacies, 189, 219–23
racism: addressing in writing studies programs, 222; computer coding, 220; imposter syndrome, 178; linguistic, 29, 217, 218, 223n1; negotiation tools, 54; social constructs, 201; students of color, 231, 232
racist assessment practices, 72
rasquachismo, 234–35
Rayley, Rita, 220
(Re)Considering What We Know, 64
reading, love of, 196–97
Reclaiming Composition for Chicano/as and Other Ethnic Minorities: A Critical History and Pedagogy, 159, 221
relational epistemologies, 46, 57
relational equilibrium, 237n2
relevancy of first-year writing courses, 30
remedial education, 59n13
renaming writing studies programs, 7–8
research agendas, 48, 69, 222–23
research methods, 61, 71, 209, 233–34
research participants: Black female graduate students, 177; challenges of teaching, 61–62; changing attitudes for the discipline, 38–39; current trends in writing studies, 97; demographics, 59n15, 59n16; disciplinary identity, 25–28, 40–41; diversity, 59n15, 59n16; generational subsets, 58n1; lack of diversity, 59n15, 59n16; narratives, 11; origin stories, 34; personal statement, 214–15; queer methodologies, 13; retired writing scholars, 12
resilience, claiming disciplinarity, 13–14, 103, 104, 149, 152–54, 159

respectability politics, 185–86
Restaino, Jessica, 11, 13, 21, 45, 72, 76, 77, 79, 233
"Retention ≠ Panopticon: What WPAs Should Bring to the Table in Discussion of Student Success," 71
retention rates of students, 70–72
retirement, faculty, 24–26, 44
rhetmap, 11
Rhetoric and Composition as Intellectual Work, 5
rhetoric and composition studies: Black representation, 215, 218; coding literacy, 204, 220–22; collective reality, 203; curriculum reform, 29–30; digital composing, 210; disciplinarity, 8, 29, 44, 62; doctoral programs, 27; energy, 203–4; feminist theory, 198–99; flexibility, 96; graduate coursework, 68; history of, 27; instructors, 58n3, 59n9; language constructs, 204; literature/composition divide, 37, 88–90; major, 63; material interaction, 199–200, 204; multimodal composition instruction, 210; object-oriented ontology, 198–99; otherness, 46; personal identity, 31; pluralingualism, 184; potential, 203; scholarship, 29; social constructs, 203; staffing, 68; text remediation, 210; translingual writing, 216–17. *See also* writing studies; writing studies programs
rhetoric and writing studies: affect of language, 202; disciplinary, 3–4, 201; discrimination, 149; Hispanic teacher-scholars, 148; multidisciplinary, 7; relationships with writing, 197; self-identity, 25; tenure-track jobs, 3; translingual approaches, 54–55. *See also* writing studies; writing studies programs
rhetoric of race, 189
Rhetorical Feminism and This Thing Called Hope, 160n7
The Rhetorical Tradition, 149
Rich, Adrienne, 125
Richardson, Elaine, 184
Ridolfo, Jim, 11, 135
Rinderer, Reggie, 51
Rios, Gabriela Raquel, 155
Ritchie, Joy S., 153
Robertson, Liane, 66
Robinson, Subrina J., 177
Rogers, Sean, 173
Rosas, Raymond D., 15, 16, 30, 126, 134, 152, 165, 173, 232
Rose, Mike, 184

Rose, Shirley, 66
Royster, Jacqueline Jones, 4
Ruiz, Iris, 134, 159, 221

Sánchez, Raúl, 159
Sano-Franchini, Jennifer, 141, 143
SAT scores, 27
Schell, Eileen, 11
scholarly journals, 23
scholars of color: academic space, 191; affective labor, 134, 170; alienation, 159–60n1; assimilation, 160n7; belonging, sense of, 8, 10; creating space, 230–31; discrimination, 151–52, 190–91; embodied knowledge, 151–52, 157; emotional labor, 133–35, 170, 175, 186; enrollment, 140; expectations, 186; financial labor, 172–76; imposter syndrome, 160n2, 175, 231, 232; intersectionality, 152; lived experiences, 184–85; multicultural narratives, 10; otherness, 159–60n1; personal identity, 157; predominantly white institutions (PWIs), 15, 191; professional identity, 5, 15, 187, 222, 232; research agendas, 223; respectability politics, 185–86; self-advocacy, 173–74; shared academic experiences, 179–80; tokenism, 171–72; trauma narrative, 176; uncompensated labor, 133–35
scholarship: academic experiences, 178; antiracist practices, 142–43; coding literacy, 220; cross-disciplinary, 38, 63–64, 195; diversity, 232; field of composition, 29; Indigenous scholars, 201; knowledge-making, 57, 195; minoritized voices, 52; new materialism, 200–201; peer networks, 35; plurilingual work environments, 183; positionality, 195; process approach, 55–56; race and technology, 219–22; teaching of writing, 34, 44–45; tenure-track faculty, 114; writing studies, 31, 109
school integration, 59n14
screen studies, 7
self-advocacy, 173–74
self-reflection, 145, 176
semi-structured interviews, 24–25
separation process, 76, 79, 80, 99n1
service discipline, 5
service positions, 89
shared decision-making, 97–98
Shaughnessy, Mina, 51, 53, 55
Shavers, Marjorie C., 177
Shelton, Cecilia D., 143
shifting disciplinarity attitudes, 38

Silverman, David J., 58n8
slow agency, 13, 68, 72–73
Smith, Leonie, 174
Smitherman, Geneva, 4
social awareness in writing studies, 219
social constructs, 198–201, 203
"The Social Ecology of Tokenism in Higher Education," 167
social justice initiatives, 28, 33, 54, 77, 78, 99, 113–14, 157, 208
social media, 3–4, 211
social reality, 217
social science research methods, 48
social turn, 56
social uplift, 191–93
Sommers, Nancy, 56, 59n15
Sotirin, Patricia, 153
space. See academic space
special interest groups (SIGs), 36, 143–44
specializations in writing studies, 39, 40, 49, 110
disciplinary identities, 110
speech act theory, 48
standalone writing studies programs, 7
standard English, 217, 221, 223n1
standardization of curriculum development, 66–68, 96, 208
state requirements, higher education, 13, 45, 72, 73n1
static disciplinarity, 73
status of writing studies, 72
Steele, Claude, 212
stipends, 97, 135, 139, 173
storying: disciplinary identity, 12; embodied knowledge, 234; literacy, 211; lived experiences, 229; making, 234, 237n2; meaning-making, 230; storytelling, 237n2; text remediation, 211; writing studies, 16
strategy, disciplinarity, 7–8, 15, 21–22, 26
stretch programs, 28
Strickland, Donna, 132
student input, 210–11
student literacy, 29
student persistence rates, 70–72
student population, 31
student retention rates, 70–72
student success committees, 71
students of color: creative writing programs, 151; deficit thinking, 216, 217; depression, 232; digital literacy, 219; discrimination, 150, 169, 231, 232; doctoral program applications, 169–70; embodiment, 15; emotional labor, 158, 232; enrollment, 28,

52–53; financial labor, 144–45; otherness, 151–52, 231; performance expectations, 168; social ideologies in the classroom, 219; supportive spaces, 143–44
Students' Right to Their Own Language, 12, 54, 28, 217, 218
subjectivity, 211
subspecialities in writing studies programs, 39
supportive academic spaces, 142–44, 200
survey methodology, 25
systemic racism, 54

tactical positioning for disciplinarity, 26, 29, 30, 168
tactile/cultural experience, 237n2
Taczak, Kara, 66
taxonomy of research disciplines, 9, 63
teacher-scholars: becoming, 104; belonging, sense of, 8; BIPOC, 148, 168; disciplinary origin stories, 29; emotional labor, 103–4; interconnectedness, 33; job security, 59n14, 111; lived experiences, 11; positionality, 103–4; preservation of culture, 221; professional identity, 165; transitioning to writing studies, 28
Teaching Assistants' Association (TAA), 30
teaching of writing: antiracist practices, 151, 209, 223; classroom management, 219; connectedness, 56; creative writing pedagogy, 105, 111, 112, 121, 126–27; current-traditionalist, 44–45; digital technologies, 210–11; disciplinarity, 24–26, 32–33, 37, 38, 56–57, 113, 132; emotional labor, 143; first-year composition, 65–66, 113, 121; inclusion, 209; limitations of language, 205; literature/composition divide, 37, 88–90; love of, 32–33, 107; mentorships, 34; otherness, 37; pedagogical shift, 44–45; potential, 235; process research, 55–56; public education systems, 219; relational epistemologies, 57, 197–98; research writing, 61–62; scholarship, 34, 44–45; social justice, 33; writing studies programs, 76, 99n2
techné, 78, 79, 92
technical writing courses, 32, 106–7, 136, 136–37
technology and race, 204, 219–22
Tellez-Trujillo, Karen R., 14, 16, 85, 103–4, 133, 184, 232
temporality, 73, 195
tenure-track faculty: administration positions, 91; collective bargaining agreements, 109; committee memberships, 116; conversion process, 125–26; creative writing studies, 128–29; curricular committee membership, 84–85; emotional labor, 141–42; job security, 31, 110–11, 114, 116, 122, 124–26; policy negotiations, 106; professionalism, 141–42; relationship with full-time faculty, 78; retirement, 26; rhetoric and writing studies, 3; studies lines, 81; workload, 77, 110; writing studies programs, 77, 99n2. *See also* faculty; non-tenure-track faculty
terminal degrees, 58n9, 112, 115, 119, 123, 128–29
TESL/linguistic studies, 7
text remediation, 210–11
textual consumption/evaluation/production, 184
Thanksgiving deconstruction myth, 58n8
theorized examinations, 11
theory wars, 40
thing and environment, 16
3/4 teaching workload, 81
threshold concept theory, 63, 64
tipping points, 216
tokenism: assimilation, 167; emotional labor, 134, 169, 192; female scholars of color, 179–80; graduate programs, 232–33; hypervisibility, 167, 170; polarization, 167, 175–76; predominantly white institutions (PWIs), 15, 167; professional identity, 168; scholars of color, 168, 171–72, 178; social identity, 167–68; writing studies programs, 179–80
training. *See* professional development
trajectories, careers, 29, 208
transfer-based pedagogies, 63
transformation, cross-disciplinary studies, 36–37
Translingual Inheritance: Language Diversity in Early National Philadelphia, 54–55
translingual writing in composition, 216
translingualism, 216, 218
trauma narrative, 176, 229
Trimbur, John, 54–55
Turner, Lorenzo Dow, 217
tutor jobs, 28

uncompensated labor, 132–35
unconscious bias, 217
undergraduate majors, 6–7, 81, 94, 151
underprepared students, open admissions policies, 53
undisciplinarity, 65–66

union contracts: claiming writing studies, 129; collective bargaining agreements, 109; curriculum development, 30; departmental policies, 106; instructional specialist faculty, 77, 82–83; McGuire Memorandum, 110–110, 112; negotiation, 129n1; part-time faculty, 118–19; stipends, 97
universities. *See* academia; colleges and universities; higher education
Untenured Faculty as Writing Program Administrators: Institutional Practices and Politics, 58n3
upper-division writing requirements, 69, 70
upward mobility, 132, 133, 139, 192

value of discipline, 47
Vandenberg, Peter, 5
VanHaitsma, Pamela, 144
Vee, Annette, 219
vertical writing instruction, 69
Villanueva, Victor, 184, 160n6
Visibility Project, 4, 9, 62
Vitanza, Victor J., 197
Voices of the Self: A Study of Language Competence, 184
von Petersdorff, Anne, 195
Voos, 173

WAC/WID. *See* writing across the curriculum (WAC) / writing in the disciplines (WID)
Wardip-Fruin, Noah, 220
Wardle, Elizabeth, 5, 61, 64
Warner, Michael, 90, 94
wealth influence on writing studies, 235
Weiser, Irwin, 66
Wells, Susan, 31
West, Cornel, 212
West Chester University (WCU), 109–14
whiteness, 4, 14, 16, 133, 134, 151, 208, 220
white privilege, 54, 131, 132, 136
white racial habitus, 133, 208, 209–11, 217–19, 221
white supremacy, 213
Williams, Jeffrey, 135, 136, 137
Wilson, Darren, 212
Wingard, Joel, 33, 34, 35, 37
Wisconsin-Madison, University of, 30
women studies, 77
Wood, Tara, 13, 21, 41n1, 105–6, 111, 234
Woodson, Carter, 4
work/life balance, 31
workload, faculty, 78, 81, 89–92, 109–12, 119–21, 139–40, 168

workshops, professional development, 115
WPA Outcomes for First-Year Writing, 210
Writing about Writing, 61
writing across the curriculum (WAC) / writing in the disciplines (WID), 62, 69–70
Writing against the Curriculum: Anti-Disciplinarity in the Writing and Cultural Studies Classroom, 70
writing centers, 142
writing instructors. *See* teaching of writing
The Writing Program Administrator, 66
writing program administrators (WPAs), 8, 33–34, 62, 65–68, 71–73, 132
Writing Program Administrators, Council of (CWPA), 35
writing program directors (WPDs), 97
writing studies: allegiances, 165; BIPOC, 10, 208, 209, 221, 222; claiming, 73, 77, 105, 108, 129; content/outcomes articulations, 73n1; disciplinarity, 3–5, 13–14, 25, 26, 126; diversity, 37–38, 40–41, 98–99, 109, 140, 159; emerging field, 29, 63; equity, 230; evolution, 27, 38–39; frameworks, 214, 222–23; future of, 233–34; history of, 12, 27, 23; individual narratives, 41; invisibility, 13; language, power of, 15–16, 195–97, 205; legitimacy, 35, 45; meaning-making, 219, 234–35; ontological approach, 197–98; origins, 7, 8, 23; otherness, 58n4; pathways into, 142, 144–45; pedagogy versus theory, 5, 40–41; positionality, 158, 233, 235; process research, 55–56; professional identity, 7–8, 15, 23–24, 132; queer theory, 233; racial identity, 5, 58n4; respectability politics, 185; scholarship, 23, 31; self-reflection, 145; social justice, 54, 77, 78, 99, 157, 211, 219; theory wars, 40; tipping points, 216. *See also* rhetoric and composition studies; rhetoric and writing studies
writing studies programs: administrative roles, 33–34; cross-disciplinary learning, 5–6, 64, 77, 218, 234; curricular development, 13, 21–23, 69–70, 73n1, 93–96, 99n1; department formation, 72, 76, 79–81, 86–87, 106; doctoral programs, 59n10; exclusion, 158–59; English majors, 207–8; faculty, 25, 76, 58n2, 58n9, 81, 97, 99n2; first-year writing, 70, 107–8; flexibility, 59n11; foundation grants, 29; funding, 45; gender identity, 58n4; graduate programs, 9, 59n10, 99n1; independent departments, 58n5, 58n6, 58n7; job security, 106–7, 111; labor conditions, 105; literary studies,

32–33, 44–46; literature divide, 81, 88–90, 112–13; lived experiences, 185; pedagogy, 197; policy changes, 105–6; renaming, 7–8; specializations, 39, 40, 110; technical writing, 106–7; terminal degrees, 58n9; tokenism, 179–80; training, 38, 109; undergraduate majors, 6–7, 63, 81, 94. *See also* rhetoric and composition studies; rhetoric and writing studies

written narratives, African American literary tradition, 212–13

Yancey, Kathleen Blake, 40–41, 45, 51, 55, 56, 57, 63, 210
Young, Vershawn Ashanti, 5, 143, 168, 171, 216

Zamin, Nadia Francine, 134, 140
Zepeda, Alison Wells, 15–16, 165, 235

About the Authors

Khadeidra (Khay) Billingsley is a third-year PhD candidate in English specializing in composition, rhetoric, and English studies at The University of Alabama. She currently holds a BA in English from Clark Atlanta University, an MA in English from DePaul University, and an MA in educational psychology. Khay's background in educational psychology and English greatly informs her current scholarship as her dissertation focuses on illuminating high school English teachers' perceptions of college writing. Her research interests include composition pedagogy, K–12/college collaboration, and African American rhetoric. Her work has been published in several edited collections and *Kairos* journal. Khay has received several teaching awards at The University of Alabama and is a 2020 National Council of Teachers of English (NCTE) Early Career Educator of Color Leadership Award recipient. She is also a Mellon Mays UNCF fellow as well as a Southern Regional Education Board (SREB) doctoral fellow.

Lauren Marshall Bowen is an associate professor of English and director of the Composition Program at the University of Massachusetts, Boston, where she teaches writing, literacy, composition theory, and composition pedagogy. Her research interests include literacy, age studies, writing through the lifespan, and composition pedagogy. Her research on age inclusivity in higher education has been funded by the RRF Foundation for Aging. Her work has appeared in

College Composition and Communication, College English, Community Literacy Journal, Literacy in Composition Studies, Computers and Composition, Gerontology and Geriatrics Education, and various edited collections.

Ron Brooks is an associate professor and founding chair of the Department of Writing Studies at Montclair State University. He teaches courses in writing, rhetoric, and writing center pedagogy. He has published in the journals *College Composition and Communication, Technical Communication Quarterly, Enculturation, Pretext*, and has co-written articles on writing center and composition program collaborations.

Antonio Byrd is an assistant professor of English at the University of Missouri-Kansas City, where he teaches courses in Black/African American literacies, professional and technical writing, multimodal composition, and digital rhetoric. His research focuses on computer programming as a literacy; his current book project studies how Black adults attending a computer code bootcamp access, learn, and use computer programming to address racial inequality. His work has appeared in *College Composition and Communication* and *Literacy in Composition Studies*.

Christina V. Cedillo is an associate professor of writing and rhetoric at the University of Houston-Clear Lake where Christina teaches first-year and advanced writing courses. Her/their research examines embodied rhetorics and rhetorics of embodiment at the intersections of race, gender, and disability and highlights these identities in the creation of critical inclusive pedagogies. Drawing on critical race theory, disability rhetorics, and decolonial theories, Christina's research highlights rhetorical tropes and topics across time periods that expose how colonization and coloniality affect our lives. Her/their work has appeared in *Composition Studies, Rhetoric Society Quarterly, College Composition and Communication, Feminist Studies in Religion, Argumentation and Advocacy, Present Tense, Composition Forum*, the *Journal for the History of Rhetoric*, and various edited collections.

Caroline Dadas is an associate professor of writing studies at Montclair State University, where she teaches courses in writing, rhetoric, and gender studies. She has published in the journals *College Composition and Communication, Computers and Composition, Peitho, Literacy in Composition Studies*, and *New Media and Society*. She is co-editor (with William Banks and Matthew Cox) of *Re/Orienting Writing Studies: Queer Methods, Queer Projects* (2019).

Suellynn Duffey's long career began during the late-twentieth-century resurgence of composition and rhetoric studies. Thus, her professional life carries traces of the discipline's (recent) history. Currently, she directs the writing programs at the University of Missouri-St. Louis and has done so previously at The Ohio State University, the University of Wisconsin-Eau Claire, and Georgia Southern University. She has published in *CCC*, *Writing on the Edge*, *Rhetoric Review*, *The Journal of Public Scholarship in Higher Education*, and *Writing Program Administration: Journal of the Council of Writing Program Administrators*. She has contributed chapters to several books on a variety of subjects including pedagogy, WPA work, silence and listening, and women's professional lives in rhetoric and composition. Her current scholarly project is a book on place and literacy intersections.

Alison Ersheid teaches composition, along with courses in business and organizational writing. Her research focuses on culturally relevant pedagogy and practices that support minoritized students.

Laura Field, PhD, is a full-time non-tenure-track instructor in the Writing Studies Department at Montclair State University where she teaches classes in writing and rhetoric. In addition to teaching, her time and efforts are focused on labor issues and advocating for contingent faculty.

Cynthia Johnson is an assistant professor of English at the University of Central Oklahoma. Her research examines the tools, technologies, and power structures that mediate learning transfer. Her work has appeared in *The Writing Center Journal*, *Computers and Composition*, and the digital edited collection *Transfer of Learning in the Writing Center*.

Seth Kahn teaches composition along with courses in activist rhetoric, propaganda, and qualitative research methods. His scholarship focuses on academic labor activism. Recent publications include "We Value Teaching Too Much to Keep Devaluing It" (*College English*, July 2020) and *Activism and Rhetoric*, 2nd edition, co-edited with JongHwa Lee (Routledge, December 2019).

Lisa Konigsberg teaches composition, business writing, and research writing. She has published two collections of poetry. *Invisible Histories*, Spruce Alley Press, 2015, and *The Golden Mean*, Moonstone Publishing, 2019. She has presented research at the Popular Culture Association in Men and Masculinities on the topic of gun violence and is working on a paper currently titled "Not Just Talking Heads: Creating Community in Virtual Classrooms."

Maureen McVeigh teaches composition and creative writing. Her essays and short stories have appeared in *Sonder Midwest*, *Cold Noon*, *Mothers Always Write*, *Flash Fiction Magazine*, and *Calyx*. "Experience and Imagination: A Pedagogical Approach to Write What You Know" was recently published in *New Writing: The International Journal for Creative Writing*.

Nancy Pearson teaches composition and creative writing. Her publications include the collections of poems *The Whole by Contemplation of a Single Bone* (Fordham University Press) and *Two Minutes of Light* (Perugia Press). Her writing has garnered two seven-month fellowships at the Fine Arts Work Center in Provincetown and a PEN/New England Award.

Laurie A. Pinkert is an associate professor of writing and rhetoric and director of Writing Across the Curriculum at the University of Central Florida. Her scholarship has appeared in *College Composition and Communication*, *College English*, *Composition Studies*, the proceedings of the American Society for Engineering Education, and *Reflections: A Journal of Community-Engaged Writing and Rhetoric*; and *WPA: Writing Program Administration*. Her research on disciplinary identity development within and through writing, which extends beyond the field of writing studies, has been funded by the National Science Foundation to further investigate the relationships between individual and collective values, STEM enculturation, and disciplinary identification.

Jessica Restaino is a professor of writing studies and director of Gender, Sexuality, and Women's Studies at Montclair State University. She is the author of *First Semester: Graduate Students, Teaching Writing, and the Challenge of Middle Ground* (2012); co-editor (with Laurie Cella) of *Unsustainable: Re-Imagining Community Literacy, Public Writing, Service-Learning, and the University* (2012); and author of *Surrender: Feminist Rhetoric and Ethics in Love and Illness* (2019).

Raymond D. Rosas is a PhD candidate in rhetoric and writing studies at Penn State. His major academic interests include critical literacy studies, genre systems theory, and the rhetoric of health and medicine. An advocate of the teacher-scholar model of work in the academy, he believes one's research and pedagogical commitments should function symbiotically toward innovation and professional excellence. In 2019, he received an excellence in teaching award from Penn State's program in writing and rhetoric. His dissertation work examines the politics of writing instruction, especially as they pertain to African American and Latinx students in predominantly white institutions.

Karen R. Tellez-Trujillo is an assistant professor at Cal Poly Pomona in the Department of English and Modern Languages with an educational background in border rhetorics, gender and sexuality studies, and rhetoric and composition. As a graduate of New Mexico State University with a Bachelor of Arts in English and in women's studies, an MA in rhetoric and professional communication, and a PhD in rhetoric and professional communication, she contributes to her department in unique ways that reflect her many intersections. She comes to teaching with the experience of being a teen mother, a young business owner, a woman with a chronic illness, a Chicanx feminist, and a lifelong resident of the Borderlands in which she is nepantla (Anzaldua), residing in many locations at once. Karen has been involved in over ten publications and twenty-three presentations for research, which have been recognized internationally. Her dissertation, "Enactments of Feminist Resilience: Rescripting Post-Adversity Encounters through Pause and Reflection," is a qualitative study of how students respond to adversity with resilience. While at Cal Poly, she looks forward to teaching courses in which students can focus on resilience and the way it is expressed in the composition classroom.

Tara Wood is an associate professor of English and a writing program administrator at the University of Northern Colorado. Her research interests include disability, writing pedagogy, and writing program administration. Her work has appeared in several essay collections and journals, including *College Composition and Communication*, *Composition Studies*, and *Open Words: Access and English Studies*. Her scholarship has been honorably recognized by the Committee on Disability Issues in College Composition, by the Coalition of Feminist Scholars in Rhetoric and Composition, and by Computers and Composition Digital Press. In addition to her scholarship, Dr. Wood has also held two elected leadership positions within CCCC, serving as past co-chair of the Committee on Disability and as a former member of the Executive Committee.

Alison Wells Zepeda is a PhD candidate in rhetoric and writing studies at The University of Texas at El Paso. She teaches dual credit, dual enrollment, and advanced placement courses in English at Jefferson-Silva Health Magnet High School in El Paso, TX. Her research interests are focused on the convergences of rhetoric, affect, and the posthuman and the ethical considerations of these theoretical concepts in the teaching of writing and composition. Alison is co-author of "International Writing Tutors Leveraging Linguistic Diversity at a Hispanic-Serving Institution's Writing Center" for *The Peer Review*, a journal of the International Writing Center Association. She is currently reviewing *Rhetoric in Tooth and Claw: Animals, Language, Sensation* by Debra Hawhee for *Present Tense: A Journal of Rhetoric in Society*.

www.ingramcontent.com/pod-product-compliance
Lightning Source LLC
Chambersburg PA
CBHW060555080526
44585CB00013B/568